Statistical Process Control for the FDA-Regulated Industry

Also available from ASQ Quality Press:

Quality Risk Management in the FDA-Regulated Industry
José Rodríguez-Pérez

The FDA and Worldwide Quality System Requirements Guidebook for Medical Devices, Second Edition
Amiram Daniel and Ed Kimmelman

CAPA for the FDA-Regulated Industry
José Rodríguez-Pérez

Development of FDA-Regulated Medical Products: A Translational Approach, Second Edition
Elaine Whitmore

Medical Device Design and Regulation
Carl T. DeMarco

The Quality Toolbox, Second Edition
Nancy R. Tague

The Certified Six Sigma Green Belt Handbook
Roderick A. Munro, Matthew J. Maio, Mohamed B. Nawaz, Govindarajan Ramu, and Daniel J. Zrymiak

The Certified Manager of Quality/Organizational Excellence Handbook, Third Edition
Russell T. Westcott, editor

The Certified Six Sigma Black Belt Handbook, Second Edition
T. M. Kubiak and Donald W. Benbow

The ASQ Auditing Handbook, Fourth Edition
J.P. Russell, editor

The Internal Auditing Pocket Guide: Preparing, Performing, Reporting, and Follow-Up, Second Edition
J.P. Russell

Root Cause Analysis: Simplified Tools and Techniques, Second Edition
Bjørn Andersen and Tom Fagerhaug

Statistical Process Control for the FDA-Regulated Industry

Manuel E. Peña-Rodríguez

ASQ Quality Press
Milwaukee, Wisconsin

American Society for Quality, Quality Press, Milwaukee 53203

Printed in the United States of America
19 18 17 16 15 14 5 4 3 2

Library of Congress Cataloging-in-Publication Data

Pena-Rodriguez, Manuel E.
Statistical process control for the FDA-regulated industry / Manuel E. Pena-Rodriguez.
pages cm
Includes bibliographical references and index.
ISBN 978-0-87389-852-2 (hardcover : alk. paper)
1. Process control—Statistical methods. 2. Manufacturing processes—United States—Quality control. I. Title.

TS156.8.P45 2013
658.5072'7—dc23 2013003376

ISBN: 978-0-87389-852-2

Publisher: William A. Tony
Acquisitions Editor: Matt T. Meinholz
Project Editor: Paul Daniel O'Mara
Production Administrator: Randall Benson

Printed on acid-free paper

ASQ

Quality Press
600 N. Plankinton Ave.
Milwaukee, WI 53203-2914
E-mail: authors@asq.org

The Global Voice of Quality™

To my daughter Stacey Marie and to my mom Sonia Rodríguez. Thanks for always being my inspiration in everything I do.

And to my best friend José (Pepe) Rodríguez-Pérez. Thanks for believing in me and helping me to make this dream come true.

Table of Contents

List of Figures and Tables

Preface

Over the centuries, the *quality* of products and services has been one of the common characteristics of successful organizations. The term "quality" has evolved through the generations. Philosophies such as *quality control*, *quality assurance*, and *total quality management* have been recognized at different times. Nevertheless, all these philosophies share something in common: the use of *statistical process control* (SPC) to achieve higher levels of excellence. The concept of SPC applies to any type of industry: automotive, textiles, pharmaceutical, biologics, medical devices, electronics, aerospace, banking, educational services, and so on.

With the advances in technology, more people are immersed in the SPC arena every day. Computer software such as Minitab, Statgraphics, SigmaXL, and others, make the analysis of data a simpler task. However, most of the questions that people ask me every day are not about how to perform the analysis once the person determines which tool to use, but about *which* is the appropriate tool to use for each specific situation.

The focus of this book is to understand and apply the different SPC tools in a company regulated by the Food and Drug Administration (FDA): those that manufacture pharmaceutical products, biologics, medical devices, food, cosmetics, and so on. The book is not intended to provide an intensive course in statistics; instead, it is intended to provide a how-to guide about the application of the diverse array of statistical tools available to analyze and improve the processes in an organization regulated by FDA. This book is aimed at engineers, scientists, analysts, technicians, managers, supervisors, and all other professionals responsible to measure and improve the quality of their processes. Although the examples and case studies presented throughout the book are based on situations found in an organization regulated by FDA, the book can also be used to understand the application of those tools in any type of industry.

The book comprises 12 chapters and four appendixes. In Chapter 1, the regulatory importance of SPC is presented. Some of the FDA regulations and guidances are analyzed in terms of the agency's expectations about the use of statistical process control tools. Also, some of the international standards applicable to the life sciences industry are analyzed for SPC requirements. Chapter 2 presents various instances in which FDA has issued observations about the misuse of SPC tools. Also, the concepts of SPC and *corrective action and preventive action* (CAPA) are integrated in this chapter.

Then, Chapter 3 presents the concept of *process variation*. The common causes and special causes of variation are explained in detail. Chapter 4 presents some basic statistical concepts, such as types of data, sampling, descriptive statistics, the normal distribution, and so on. Next, Chapter 5 presents some of the most useful graphical tools with which to start analyzing processes. Tools such as the histogram, dot plot, box plot, Pareto diagram, and others, as applied to several FDA-regulated industries, are presented in the chapter.

In Chapter 6, one of the most important but less frequently used tools is presented: the measurement systems analysis. In this chapter, the importance of addressing measurement system variability prior to implementing any other improvement initiative is thoroughly explored. Chapter 7 presents the concept of *process capability*. Here, we study the different indices used to measure capability: C_p, C_{pk}, P_p, and P_{pk}. Then, in Chapter 8, an introduction to *hypothesis testing* is presented. Several tools used to compare means, medians, and variances are introduced for normal and nonnormal data. Many examples are provided detailing the use of these tools in an FDA-regulated organization.

Chapter 9 explains how to use regression analysis to understand the relationship between input variables and output variables. Then, Chapter 10 provides a brief introduction to *design of experiments* and its application in an FDA-regulated environment. The concepts of *full factorial* and *fractional factorial* experiments are introduced in this chapter. In Chapter 11, control charts are introduced as a tool to facilitate process control. The control charts for variable data and attribute data are presented, along with some applications. Finally, Chapter 12 presents a summary of the appropriate tools necessary to reach a state of statistical process control.

In order to visualize the difference between attribute and variable data, Appendix A shows some different tools for analyzing attribute or variable data, including control charts, probability distributions, sampling plans, and measurement instruments for each type of data. Appendix B presents many graphical and statistical tools to be used for different situations, and a reference to the section in the book in which the tool can be found. Appendix

C shows an example of the basic statistics to apply for an annual product review or management review. Finally, Appendix D shows some of the most commonly used hypothesis tests in an easy-to-understand tabular format.

By means of this book, I expect that the reader will obtain a better understanding of some of the statistical tools available to control their processes. Also, I expect the reader to be encouraged to study, with a greater level of detail, each of the statistical tools presented throughout the book. The content of this book is the result of almost 20 years of experience in the application of statistics in various industries, and the combination of my engineering and law educational backgrounds, specifically through providing consulting services to dozens of FDA-regulated organizations.

1

Regulatory Importance of Statistical Process Control

1.1 OVERVIEW

The Food and Drug Administration (FDA) is the administrative agency in the United States responsible for protecting the public health by assuring the safety, efficacy, and security of human and veterinary drugs, biological products, medical devices, our nation's food supply, cosmetics, products that emit radiation, and tobacco products. FDA is also responsible for advancing the public health by helping to speed innovations that make medicines and foods more effective, safer, and more affordable, and helping the public get the accurate, science-based information they need to use medicines and foods to improve their health.[1]

The Federal Food, Drug, and Cosmetic Act (FD&C Act) is a federal law enacted by Congress. It and other federal laws establish the legal framework within which FDA operates. The FD&C Act can be found in the United States Code (USC), which contains all general and permanent U.S. laws, beginning at 21 USC §301. FDA develops regulations based on the laws set forth in the FD&C Act or other laws under which FDA operates. FDA follows the procedures required by the Administrative Procedure Act (APA), another federal law, to issue FDA regulations. This typically involves a process known as "notice and comment rulemaking" that allows for public input on a proposed regulation before FDA issues a final regulation. FDA regulations are also federal laws, but they are not part of the FD&C Act. FDA regulations can be found in Title 21 of the Code of Federal Regulations (CFR).[2] FDA follows the procedures required by its "Good Guidance Practice" regulation to issue FDA guidance. FDA guidance describes the agency's current thinking on a regulatory issue. Guidance is not legally binding on the public or FDA. The Good Guidance Practice regulation can be found at 21 CFR 10.115.[3]

While regulations are legally enforceable, guidances do not bind the companies. However, not following a guidance might result in a misinterpretation of the regulation; consequently, not following a guidance could have regulatory consequences.

1.2 PROCESS CONTROL WITHIN THE CODE OF FEDERAL REGULATIONS

Several sections within Title 21 of the Code of Federal Regulations mention the concept of process controls. I will focus the discussion on two specific sections of Title 21: the section related to finished pharmaceutical products[4] and the section related to medical devices.[5] It is important to understand that regulations are not intended to provide a specific way to achieve process controls. Regulations provide the *minimum* requirements. For instance, the regulation for finished pharmaceutical products states that:

> The regulations in this part contain the minimum current good manufacturing practice for preparation of drug products for administration to humans or animals.[6]

The regulation for medical devices establishes that:

> . . . This part establishes basic requirements applicable to manufacturers of finished medical devices.[7]

Both regulations explicitly state that requirements established therein are the minimum that the manufacturer must accomplish; they are not intended to be a "one-size-fits-all" type of requirement. Let us start with the process controls within the regulation for finished pharmaceutical products.

1.2.1 Current Good Manufacturing Practices (21 CFR 211)

Published in 1978, the current Good Manufacturing Practices (cGMP) provide a framework to control finished pharmaceutical processes. Control over the processes is important so that the product meets standards of safety, efficacy, purity, and stability. Section 211.22 establishes the responsibilities of the *quality control unit* (QCU). This section states that:

> There shall be a quality control unit that shall have the responsibility and authority to approve or reject all components, drug product containers, closures, in-process materials, packaging material,

> labeling, and drug products, and the authority to review production records to assure that no errors have occurred or, if errors have occurred, that they have been fully investigated. The quality control unit shall be responsible for approving or rejecting drug products manufactured, processed, packed, or held under contract by another company.[8]

In order to comply with this section of the regulation, the manufacturer shall establish written procedures, which shall be followed. It should be noted that the QCU is responsible to establish all process controls, monitor those process controls, and take actions whenever those process controls are not followed. In other words, the QCU becomes the "arms and eyes" of FDA within the manufacturer. One of the best examples of this application is found in section 211.100, "Written procedures; deviations," which states that:

> There shall be written procedures for production and process control designed to assure that the drug products have the identity, strength, quality, and purity they purport or are represented to possess. Such procedures shall include all requirements in this subpart. These written procedures, including any changes, shall be drafted, reviewed, and approved by the appropriate organizational units and reviewed and approved by the quality control unit.[9]

As noted, two key elements in establishing process controls in a pharmaceutical manufacturing environment, as established by the regulations, are the appointment of a quality control unit and the development of written procedures.

1.2.2 Quality System Regulation (21 CFR 820)

Published in 1996, the current Quality System Regulation (QSR) provides a framework to control medical device processes. Although the regulation related to pharmaceutical products (21 CFR 211) does not have a specific section dedicated solely to statistical process control, the regulation related to medical devices addresses SPC explicitly. Section 820.250, "Statistical techniques," establishes that:

> (a) Where appropriate, each manufacturer shall establish and maintain procedures for identifying *valid statistical techniques* required for establishing, controlling, and verifying the acceptability of process capability and product characteristics.

> (b) Sampling plans, when used, shall be written and based on a *valid statistical rationale.* Each manufacturer shall establish and maintain procedures to ensure that sampling methods are adequate for their intended use and to ensure that when changes occur the sampling plans are reviewed. These activities shall be documented.[10]

Furthermore, section 820.100, "Corrective and preventive action," states that:

> (a) Each manufacturer shall establish and maintain procedures for implementing corrective and preventive action. The procedures shall include requirements for:
>
> (1) Analyzing processes, work operations, concessions, quality audit reports, quality records, service records, complaints, returned product, and other sources of quality data to identify existing and potential causes of nonconforming product, or other quality problems. *Appropriate statistical methodology* shall be employed where necessary to detect recurring quality problems.[11]

As may be noted, the regulation for medical devices explicitly establishes the use of statistical techniques for process control. It does not prescribe any specific statistical tool or technique, but establishes that the technique used must be "valid." Furthermore, the regulation also establishes that sampling must have a "valid statistical rationale." In both cases, "valid" means that tools used must be acceptable, reasonable, and appropriate to the situation at hand. So, the right tool must be used for each situation. That is basically one of the goals of this book: to allow the reader to identify which of the available statistical tools and techniques is the most appropriate for each situation. Also, the regulation for medical devices has a section that defines the corrective and preventive action process. It establishes the importance of analyzing data to identify existing and potential sources of nonconforming product. This could be achieved by the use of the appropriate statistical tools and techniques.

In summary, the regulations for finished pharmaceutical products and medical devices establish the need to control the processes. The regulation for finished pharmaceutical products does not have a specific section for statistical techniques as the regulation for medical devices does. However, it is important to recognize that, although there is a regulation for finished pharmaceutical products and a regulation for medical devices, they complement each other. That is, we need to look at both from a holistic point of view. Let us turn our attention to another set of documents published by FDA: the guidances.

1.3 PROCESS CONTROL WITHIN THE FDA GUIDANCES

As mentioned in Section 1.1, regulations are legally enforceable requirements, while guidances represent the agency's current thinking on a certain topic. However, not following a guidance might result in a misinterpretation of the regulation, which, in turn, can carry regulatory consequences. FDA frequently publishes guidances to clarify gaps in the regulations. There are many guidances that consider process controls. However, we will focus our attention to three of those guidances:

- *Quality System Approach to Pharmaceutical Current Good Manufacturing Practices*
- *Investigating Out-of-Specification (OOS) Test Results for Phamaceutical Production*
- *Process Validation: General Principles and Practices*

Our goal is to understand some of the requirements pertaining to process control in those guidances and to encourage the reader to look at the appropriate guidances published by FDA. By studying and applying the guidances to our processes, we can fill the gaps produced by a misinterpretation of the regulations.

1.3.1 Quality System Approach to Pharmaceutical cGMP Regulations

In 2006, FDA published a guidance titled *Quality System Approach to Pharmaceutical Current Good Manufacturing Practices*. In one section of the guidance, FDA establishes that:

> Under a quality system, trends should be continually identified and evaluated. One way of accomplishing this is the use of *statistical process control*. The information from trend analyses can be used to continually monitor quality, identify potential variances before they become problems, bolster data already collected for the annual review, and facilitate improvement throughout the product life cycle. Process capability assessment can serve as a basis for determining the need for changes that can result in process improvements and efficiency.[12]

The guidance recommends the use of statistical process control for process monitoring on a continuous basis. It is interesting to note that terms such as "trend analysis," "potential variances," and "process capability" are mentioned in the guidance. Those terms, and their application to an FDA-regulated organization, will be discussed in greater detail throughout the upcoming chapters of this book.

1.3.2 Investigating Out-of-Specification (OOS) Test Results for Pharmaceutical Production

An interesting application of the use of averages is presented in the 2006 FDA Guidance titled *Investigating Out-of-Specification (OOS) Test Results for Pharmaceutical Production.* Section C.1.a establishes that:

> Averaging data can be a valid approach, but its use depends upon the sample and its purpose. For example, in an optical rotation test, several discrete measurements are averaged to determine the optical rotation for a sample, and this average is reported as the test result. If the sample can be assumed to be homogeneous, (i.e., an individual sample preparation designed to be homogenous), using averages can provide a more accurate result. In the case of microbiological assays, the U.S. Pharmacopeia (USP) prefers the use of averages because of the innate variability of the biological test system.[13]

The use of averages as a measure of central tendency will be covered later in the book, as well as other measures such as the median and the mode. The advantages and disadvantages of using the average depending on the shape of the distribution will also be discussed in Chapter 4.

1.3.3 Process Validation: General Principles and Practices

In January 2011, FDA published a new guidance about process validation. Throughout the guidance it is established that process validation for drugs (finished pharmaceuticals and components) is a legally enforceable requirement under section 501(a)(2)(b) of the FD&C Act (21 USC 351), which states the following:

> A drug shall be deemed to be adulterated if the methods used in, or the facilities or controls used for, its manufacture, processing, packing, or holding do not conform to or are not operated or administered in conformity with current good manufacturing practice to assure that such drug meets the requirements of this Act as to safety and has the identity and strength, and meets the

> quality and purity characteristics, which it purports or is represented to possess.[14]

The process validation guidance establishes that process knowledge and understanding are the basis for establishing an approach to process control for each unit operation and the process overall. Strategies for process control can be designed to reduce input variation, adjust for input variation during manufacturing (and so reduce its impact on the output), or combine both approaches. The guidance states that:

> Process controls address variability to assure quality of the product. Controls can consist of material analysis and equipment monitoring at significant processing points [§211.110(c)]. Decisions regarding the type and extent of process controls can be aided by earlier risk assessments, then enhanced and improved as process experience is gained. FDA expects controls to include both examination of material quality and equipment monitoring.[15]

1.4 PROCESS CONTROL WITHIN INTERNATIONAL GUIDANCES AND STANDARDS

1.4.1 ICH Q10

The International Conference on Harmonization of Technical Requirements for Registration of Pharmaceuticals for Human Use (ICH) is a unique project that brings together the regulatory authorities of Europe, Japan, and the United States, and experts from the pharmaceutical industry in the three regions, to discuss scientific and technical aspects of product authorization. Their purpose is to make recommendations on ways to achieve greater harmonization in the interpretation and application of technical guidelines and requirements for product authorization.[16] ICH Q10 "Pharmaceutical Quality System" was adopted in June 2008. It describes one comprehensive approach to an effective pharmaceutical quality system that is based on ISO concepts, includes applicable good manufacturing practice (GMP) regulations, and complements ICH Q8 "Pharmaceutical Development" and ICH Q9 "Quality Risk Management." ICH Q10 is a model for a pharmaceutical quality system that can be implemented throughout the different stages of a product life cycle. Much of the content of ICH Q10 applicable to manufacturing sites is currently specified by regional GMP requirements. ICH Q10 is not intended to create any new expectations beyond current regulatory requirements. Consequently, the content of ICH Q10 that is additional to current GMP requirements is optional.

ICH Q10 establishes that:

> Pharmaceutical companies should plan and execute a system for the monitoring of process performance and product quality to ensure a state of control is maintained. An effective monitoring system provides assurance of the continued capability of processes and controls to meet product quality and to identify areas for continual improvement.[17]

1.4.2 ISO 13485:2003 Standard

Main non–United States regulations (for example, European Community, Canada, and Japan) for medical devices are basically harmonized with the ISO 13485:2003 standard *Medical devices—Quality management systems—Requirements for regulatory purposes.* Canada has adopted ISO 13485:2003 as a Canadian national standard (CAN/CSA-ISO 13485:2003); in Europe, it has been adopted as EN ISO 13485:2003. The use of EN ISO 13485:2003 is not mandatory as a quality system standard, but any required system must be equivalent to or better than EN ISO 13485:2003.[18]

Section 8.4 (Analysis of data) of ISO 13485:2003, which is similar to 21 CFR §820.250 previously mentioned, establishes that:

> The organization shall establish documented procedures to determine, collect and analyze appropriate data to demonstrate the suitability and effectiveness of the quality management system and to evaluate if improvement of the effectiveness of the quality management system can be made. This shall include data generated as a result of monitoring and measurement and from other relevant sources. The analysis of data shall provide information relating to: (a) feedback; (b) conformity to product requirements; (c) characteristics and trends of processes and products including opportunities for preventive action; and (d) suppliers.[19]

It can be noted that section 8.4 of ISO 13485:2003 is a perfect complement to the subparts about statistical techniques and about corrective and preventive actions within 21 CFR 820.

1.5 SUMMARY

FDA is the administrative agency responsible to regulate manufacturers of finished pharmaceutical products, food, medical devices, tobacco, cosmetics, and other products that might have an impact on the health of the American population. There are many documents FDA publishes in order to achieve that goal; some of the most important ones are the regulations and the guidances. While regulations are legally binding, guidances are just the current thinking of the agency on certain topics. However, not following a guidance can result in a misinterpretation of the regulation, causing legal consequences. FDA also relies on some international regulatory bodies for offshore operations. Some of these regulators assess the adequacy of the quality systems using international standards, such as ISO 13485:2003.

As can be seen throughout all these documents (regulations, guidances, international standards, and so on), process monitoring is an important element of any quality system. The tools presented in this book will assist organizations in monitoring quality on a continuous basis to improve their performance. The next chapter presents some observations issued by FDA to organizations about the application or misapplication of SPC tools.

2

SPC and the Life Sciences Regulated Industry

2.1 OVERVIEW

SPC tools can be used in any environment where a process needs to be monitored in order to be improved. In an FDA-regulated industry, monitoring processes in a continuous way is of paramount importance because the product in this kind of industry impacts human health. Quality can not be inspected into a product; it must be built into the product. Concerned about the lack of use (in some cases) and misuse (in other cases) of statistical tools, FDA is compelling companies to embrace a quality system approach within their manufacturing environment.

2.2 RECENT OBSERVATIONS ABOUT MISUSE OF STATISTICAL PROCESS CONTROL

Performing a search on FDA's Warning Letters web page using the term "statistical control," we find that more than 190 Warning Letters in which that term was mentioned have been issued during the past 10 years. Although most of them have been issued to medical device companies (specifically because of the existence of section 820.250, "Statistical techniques"), there are still many Warning Letters issued to pharmaceutical sites. What follows are excerpts of some of those Warning Letters where observations about misuse of statitistical tools have been issued.

In August 2012, a Warning Letter was issued to a pharmaceutical company in Illinois. During its inspection, FDA found that no procedure was established to determine what is considered a trend. Specifically, the Warning Letter states in observation #6:

> Failure to adequately analyze service reports with appropriate statistical methodology in accordance with 21 CFR 820.100, as required by 21 CFR 820.200(b). Specifically, your procedure entitled "Servicing" [SOP048, Rev. (b)(4)], fails to provide a process for analyzing service reports based on a defined statistical methodology to identify quality issues or trends.
>
> We have reviewed your firm's response and concluded it is not adequate. Your firm did not provide details or documents defining a process for trending and whether the existing servicing procedure would be updated. Further, your firm did not provide plans to retrospectively analyze existing service reports to ensure they do not represent quality issue trends requiring attention; and, your firm did not identify actions it would take when quality trends are identified.[1]

On April 27, 2012, a Warning Letter was issued to a pharmaceutical company in Mexico. During the inspection, FDA found that no procedure was established to monitor the performance of the critical inputs that impact the variability of the product. Specifically, the Warning Letter states in observation #2:

> Your firm has not established written procedures to monitor the output and to validate the performance of those manufacturing processes that may be responsible for causing variability in the characteristics of in-process material and the drug product [21 C.F.R. §211.110(a)].
>
> For example, you did not perform adequate in-process tests for the Children's XL-3 Chewable Tablets after the (b)(4) of the production runs for lots # A136, A137, and A138 to assure the process remains in a state of control throughout the run. In addition, you failed to follow your procedure INS-05-01, entitled "Controles en Proceso" section 8.1.2.2, approved on January 21, 2010, which requires (b)(4) samples be taken at the (b)(4) of the (b)(4) operation.
>
> In your response to this letter please include provisions for frequent sampling of tablets throughout the (b)(4) operation, and include a strong scientific and quality assurance rationale for your in-process sampling and testing approach (e.g., including an understanding of variability in the process and use of suitable statistical procedures). In addition, please review all in-process test results for products distributed to the U.S. within expiry, and provide a risk assessment on all lots for which your firm can provide

> no evidence that you performed adequate in-process tests and obtained results that support release of your drug products.[2]

An observation for not complying with 21 CFR 820.250 was also made by FDA to a medical device company in Canada on December 20, 2011. In this case, the observation was about not using a statistical rationale for certain procedures. Observation #4 of the Warning Letter stated:

> Failure to adequately establish and maintain procedures for identifying valid statistical techniques required for establishing, controlling, and verifying the acceptability of process capability and product characteristics, as required by 21 CFR 820.250(a). For example:
>
> (a) Your firm provided WP 5 QC Procedures for ELISA & CLIA Kits and Components Version 3.0, which requires (b)(4) for final acceptance testing for kit release. The determination of (b)(4) was not based on a statistical rationale.
>
> (b) The (b)(4) is also described in WP 5 QC Procedures for ELISA & CLIA Kits and Components Version 3.0 and it is not based on statistical rationale.
>
> (c) The design project for the PRA test kits (7.3.6 Design and Development Validation File Name: 7.3.6-v2.doc Version 2.0 Effective: 1/8/2010 and WP 4 General Protocol for R&D of ELISA systems File Name: WP 4-v2.doc Version 2.0 Effective: 1/8/2010) did not include a statistical rationale for the verification and validation testing by the firm.[3]

Sampling is one of the most questionable items whenever FDA issues an observation about the misuse of statistical tools. During the search within FDA's website, I found some Warning Letter observations due to inappropriate sampling. For example, an observation within the Warning Letter dated January 14, 2010, issued to a pharmaceutical company in New Jersey, stated:

> In addition, Section 5.4.2, Sampling Requirements, in your Process Validation Protocol, PVP-2000M-122T-04, states that (b)(4) tablets should be collected at (b)(4) for analytical testing. However, 10 tablets were collected from 14 sampling locations for a total of 140 tablets in lot S0908003. Your response does not address this apparent deviation from your protocol. Also, be advised that the degree of validation sampling (e.g., number and frequency) and testing should be more extensive (than routine production) in order to provide sufficient statistical confidence of quality within

> a batch and between batches. Please address your confidence level when sampling a total of 140 tablets from a lot of (b)(4) tablets (protocol batch size).[4]

Another observation related to sampling techniques was issued to a medical device company on October 10, 2007. Observation #15 stated:

> Failure to establish and maintain procedures to ensure that sampling methods are adequate for their intended use, to ensure that when changes occur the sampling plans are reviewed, and that sampling plans are written based on a valid statistical rationale, as required by 21 CFR 820.250(b): For example:
>
> a. Your Asheboro, NC, facility did not have a robust endotoxin testing program for Central Venous Catheters (CVC) kits; CVC/ Dialysis Large-bore ChlorPrep Drape Chlorhexidine (CDC) kits, epidural kits, and Percutaneous Sheath Introducer kits which are all labeled "non-pyrogenic." Of approximately 30 sterilizer loads per week, only three (3) 10 tests were performed on a weekly basis. Statistical techniques were not used for control purposes where statistical techniques were applicable.
>
> b. For your Everett, MA, facility, post-sterilization functional testing after the rework of catheter [*redacted*] was not conducted on a sample from the same lot, but rather on a sample from a different catheter lot in the same sterilization load. That test resulted in the functional release of product from lot [*redacted*] which was subsequently shipped to customers. Lot [*redacted*] had been reworked due to a previous failed post-sterilization functional check that identified a hole in the catheter.[5]

Yet another observation related to sampling techniques was issued to a pharmaceutical company in Ohio in a Warning Letter dated July 13, 2004. Observation #15 of the Warning Letter stated:

> Written procedures for sampling and testing plans are not followed for each drug product [21 CFR 211.165(c)]. Specifically: The number of vials to be tested for sterility was to follow the requirements in the USP. There were instances where fewer vials were sent to the testing laboratory than were indicated in the USP.[6]

One of the areas of the regulation where the use of statistical tools is a key element is 21 CFR 820.100, "Corrective and preventive action." On June 21, 2010, a Warning Letter was issued to a medical device company in California. Observation #2 stated:

> Failure to implement procedures for implementing corrective and preventive actions, as required by 21 CFR 820.100(a). Specifically, a Corrective and Preventive Action (CAPA), either a Low Level CAPA (LLCAPA) or a High Level CAPA (HLCAPA), was not initiated per procedure number POL-016—Corrective and Preventive Action System (Revision E) for the Solanas Set Screw and Instruments whereby changes from the original (b)(4) shape to the new (b)(4) shape were implemented. CAPA report #08-016 was initiated for complaints associated with the (b)(4)'. The investigator was told by the Senior Director, Quality Control that CAPA report #08-016 extended to the Solanas family of implants but there was no documented statement in the CAPA that it included the Solanas family of implants.
>
> We have reviewed your response and have concluded that it is inadequate. You revised procedure number POL-016—Corrective and Preventive Action System (Revision F) to include the requirement that a CAPA be initiated when, "More than 3 Complaints or NCMR's for the same issue for the same product within a one month period" and the "Greater than 2-sigma increase in the rate of Complaints or NCMRs during any trend review." However, this procedure does not include a valid statistical rationale for this trend identification method.[7]

The integration of SPC and CAPA is the topic of the next subchapter.

2.3 SPC AND CAPA

It has been noted that SPC is a key element of the CAPA subsystem. Section 820.100 of the regulation related to medical devices states that:

> (a) Each manufacturer shall establish and maintain procedures for implementing corrective and preventive action. The procedures shall include requirements for:
>
> (1) Analyzing processes, work operations, concessions, quality audit reports, quality records, service records, complaints, returned product, and other sources of quality data to identify existing and potential causes of nonconforming product, or other quality problems. Appropriate statistical methodology shall be employed where necessary to detect recurring quality problems;[8]

Section 820.100(a)(1) provides a non-exhaustive list of areas where SPC might be applied. However, the most important part of this section is where it states "*to identify existing and potential causes* of nonconforming product" [emphasis added]. That wording is the basis for continuous process monitoring. It not only requests the use of process monitoring as a reactive way of controlling the process (existing causes) but also establishes the need to be more proactive when controlling the process (potential causes). As will be seen in Chapter 11, through the use of control charts we will be able to identify potential problems before they actually occur.

As we have noted, section 820.100 provides the rationale to combine the statistical process control tools with the CAPA subsystem in the medical device industry. But what about a pharmaceutical company? Is there a CAPA section within 21 CFR 211? Not specifically. However, as mentioned, FDA publishes guidances from time to time to clarify certain gaps in the regulations. Specifically, to address the CAPA issue in a pharmeceutical environment, FDA issued the 2006 guidance *Investigating Out-of-Specification (OOS) Test Results for Pharmaceutical Production.* Footnote 7 states:

> Please note that §211.192 requires a thorough investigation of any discrepancy, including documentation of conclusions and follow-up. Implicit in this requirement for investigation is the need to implement corrective and preventive actions. Corrective and preventive action is consistent with the FDA's requirements under 21 CFR part 820, subpart J, pertaining to medical devices, as well as the 2004 draft guidance entitled *Quality Systems Approach to Pharmaceutical Current Good Manufacturing Practice Regulations*, which, when finalized, will represent the Agency's current thinking on this topic.[9]

So, the same link between SPC and CAPA can be established for a pharmaceutical company. As mentioned in Section 1.2.1, regulations for finished pharmaceutical products and for medical devices complement each other. In order to implement the regulations effectively, we need to visualize them from a holistic point of view.

2.4 SUMMARY

It is evident that SPC plays an important role in any organization. In this chapter we were able to see many examples of the use (and misuse) of SPC in organizations regulated by FDA. The problem is not only present in domestic sites; we were able to see that the same issues are happening offshore. Remember that many offshore medical device companies are using ISO 13485:2003 as their quality system standard, which is very consistent with the requirements of 21 CFR 820. Furthermore, many pharmaceutical companies are using ICH Q10, which FDA has also adopted as a guidance. So, the importance of SPC in monitoring and continuously improving processes is inherent in every company regulated by FDA, whether domestic or offshore. In order to learn how to implement an effective SPC system, the concept of process variation must be well understood.

3

Process Variation

3.1 OVERVIEW

Variation is an inherent part of every process. Usually, we measure the accepted variation, considering only how much the process varies when compared to the customer specifications. Figure 3.1 shows two interpretations of this variation. The diagram on the left of Figure 3.1 shows that as long as the process is within the customer specification limits, we do not have any monetary losses. Once the process gets outside the customer specifications, then we begin to accrue monetary losses.

However, as per quality guru Genichi Taguchi, monetary losses start as soon as our process starts to shift away from the target value.[1] Furthermore, Taguchi mentioned that those monetary losses were experienced by society. As we move farther away from the target value, the monetary

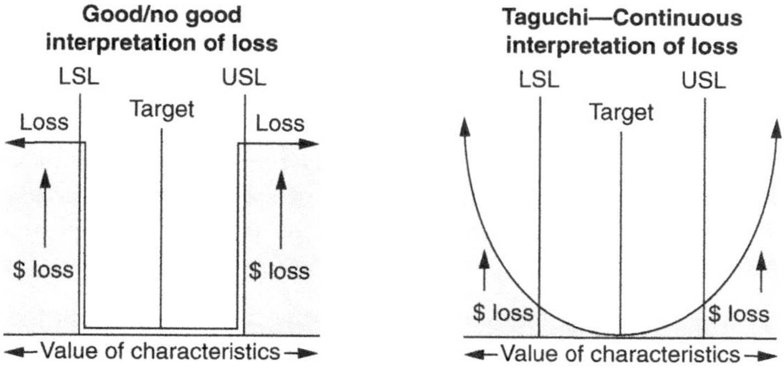

Figure 3.1 Concepts of process variation as compared to customer specifications.

losses increase, following a quadratic function. Specifically, the *Taguchi loss function* shown on the right side of Figure 3.1 is defined by the following formula:

$$L = k(y - T)^2$$

where L is the monetary loss, k is a cost factor, y is the actual value, and T is the target value.

Let us explain the concept with an example:

> A company is dedicated to the bottling of soft drinks. Their engineering department sets a target value and tolerances for the amount of soft drink that each bottle must contain. If they only focus on being within the specification limits (pass or fail decision), they will never learn about the monetary loss incurred whenever a bottle deviates from its target value (left-hand chart of Figure 3.1). In contrast, if they use the Taguchi model (right-hand chart of Figure 3.1), they can conclude that society will experience a monetary loss. How? Let us see.
>
> If their process is continuously overfilling the bottles, they will be incurring an excessive use of material (soft drink). So, when the yield of material usage is calculated, they will see a negative accounting variance. That is, in order to produce certain number of bottles they would have used a certain amount of soft drink. But, because of the overfilling, the direct material cost will be higher. The profit formula is Profit = Price – Cost. So, if *cost* increases, and they want to keep the same *profit*, then *price* has to be increased. And who do you think will be impacted by the price increase? Society, of course.
>
> If their process is continuously underfilling the bottles, the accounting variance will be positive. Now cost will be lower and profit will be higher. However, what is the problem with underfilling the bottles? Dissatisfied customers. In this case, dissatisfied customers will not purchase the product anymore. This will result in lower sales and, consequently, lower profits. If the company wants to keep the same profit, what will it have to do? Increase the price. And who do you think will be impacted by the price increase? Society, again.

In this way, we can see that society will always end up paying for the process inefficiencies. Consequently, it is of the utmost importance to reduce the process variation. In order to reach that goal, we must identify what causes the variation in our processes.

3.2 THE CAUSES OF VARIATION

Quality guru W. Edwards Deming mentioned that every process has variation. It does not matter how many times we perform a task or manufacture a product; there will always be small differences. Those differences can be attributed to the *common* and *special* causes of variation.

Common causes of variation are always present in processes. This type of variation contributes a small amount to the total variation. For instance, there might be small lot-to-lot variation. The same variation can be seen operator to operator and within operators. A characteristic of the *common* causes of variation is that they are predictable. As we will see later, when a process is in statistical control, we can predict within which values the measurements of our process will be. Consequently, we can say that a process in statistical control is stable, predictable, and subject to common causes of variation.

On the other hand, there are the *special*, or *assignable*, causes of variation. As opposed to the common causes of variation, the special causes of variation are not always present in processes. Special causes appear and disappear sporadically. The special causes are not predictable; they can happen at any time and do not necessarily provide us a signal whenever they are going to appear. A process out of statistical control shows common *and* special causes of variation. This type of process is unstable and unpredictable.

An example can help us understand the concept of common and special causes of variation:

> An operator works on a molding machine. Each day, he arrives to work and starts operating his machine at about the same time. He uses the same material, from just one supplier. He also performs the machine setting and some minor maintenance tasks. Each part that comes out of the machine is not exactly the same. There are small variations part to part. However, those small variations are considered to be due to common causes, or random process variation.
>
> One day, however, the operator is absent. A replacement operator works on the machine on that day. Although they both follow the same standard operating procedure, the machine setup is more an art than a science. That is, the setup is operator dependent. The new operator does not have the same experience as the original operator. Additionally, a material from a new supplier was

approved and started to be used on that day. Suddenly, the parts begin to vary much more than usual. The new operator starts to make adjustments to the machine until, unfortunately, it wears out. Notice that many special (or assignable) causes were present. The higher variation was produced by the combination of those special causes.

Knowledge about the causes of variation is of paramount importance because each type of variation must be dealt with using a different approach, as will be seen when we cover the topic of control charts.

3.3 SUMMARY

The study of process variation is fundamental in the implementation of an effective SPC system. We must identify and distinguish between the common causes and special causes of variation because each one must be dealt with in a different way. One of the major mistakes is to treat common causes as special causes, and vice versa. This will lead us to overreacting at times, while not reacting at all in other situations. The use of statistics can help us understand the type of variation present and which type of action would be recommended to deal with that variation.

4

Basic Principles of Statistics

4.1 OVERVIEW

In order to understand the causes of variation, we will use various statistical concepts. However, prior to going deeper into statistics, we need to learn some of the most frequently used statistical terms. *Statistics* is a collection of techniques used to make decisions about a *population* based on information taken from a *sample*. The population is the total set of data, while the sample is a subset of the population. As will be seen later, we generally take samples because measuring the population can be costly and/or time-consuming.

Descriptive statistics provides information about the data under evaluation. For instance, it helps us understand the central tendency, the dispersion, and the shape of a set of data. By using descriptive statistics, we organize, summarize, and present the data in order to make decisions. Some examples of descriptive statistics are averages, medians, ranges, variances, and so on. *Inferential statistics* allows us to make predictions about the behavior of some data, using probabilities as a mean to provide a degree of certainty to the decision we are making about the data. For instance, whenever we perform a simple linear regression we obtain an equation that allows us to predict the value of a response variable (y) for certain values of the input variable (x).

We frequently find that terms such as *parameter* and *statistic* are used interchangeably. However, that is wrong. The *parameter* is the true value, while the *statistic* is an estimate of the parameter. In order to obtain a parameter, we must measure the whole population. However, as mentioned, measuring the population sometimes is impractical. For this reason, in most cases we just take a sample and estimate the parameter by calculating a statistic. Figure 4.1 shows the symbols used to define some of the most

Metric	Parameter (population)	Statistic (sample)
Average	μ	$\bar{x}$
Standard deviation	σ	s
Variance	σ^2	s^2
Size	N	n

Figure 4.1 Symbols used for some parameters and statistics.

common parameters and statistics. Notice that parameters are obtained from the population, while statistics are obtained from the sample.

4.2 TYPES OF DATA

Prior to gathering any piece of data in order to analyze it, we need to consider the type of data we have. In this way, we will be able to know which are the available tools we will be using during the analysis of such data. To simplify our discussion, data will be split into three categories: discrete (attribute) data, continuous (variable) data, and locational data. *Discrete*, or *attribute*, data are such things that can be *counted*. Also, they can be categorical or binary. Some examples of discrete data are number of defects, shift number, machine type, good/bad decision, and so on.

On the other hand, *continuous*, or *variable*, data are such things that can be *measured*. These data can be subdivided into smaller portions. Some examples are weight, speed, temperature, and so on. This type of data provides more information than discrete data. For instance, it is not the same to say that a piece has one defect as it is to say how large the defect is.

Finally, we have *locational* data, which shows where the characteristic we are looking for is located. For instance, when we know where most of the defects occur, we might be able to arrive at a solution to the problem quicker. There is not any data type better than the others. Each data type has its own purpose. For that reason, it is recommended to use a combination of the three data types to analyze the problems and look for solutions to such problems.

4.3 SAMPLING

A *sample* is a subset of a greater set, called a *population*. There are multiple reasons to analyze a sample instead of a population. Some of those reasons are cost, time, efficiency, destructive testing, and so on. The number of samples to take will depend on several factors, such as:

- Type of data (continuous or discrete)
- Purpose of data collection
- Knowledge about the population standard deviation
- The degree of confidence we want in our results (allowable risk)

The combination of these factors will determine the amount of data to be collected. For instance, continuous data (for example, a measurement value) will require fewer samples than attribute data (for example, pass/fail decision) for the same confidence level. Less data will be required for a cosmetic characteristic than for a critical characteristic. Also, the higher the variation of the data, the larger the sample size will have to be. Finally, the more confidence we want in the results, the greater the sample size. Figure 4.2 shows an example of the kind of data collection matrix that is recommended prior to starting collection of data.

Each time that we sample, we need to consider a balance between the desired precision, the cost, and the sample size. For continuous data, we can use the following formula to calculate the sample size:

$$n = [(Z_{\alpha/2})(s)/(d)]^2$$

where

Z is a constant obtained from the normal distribution table based on the allowed error

s is the estimated standard deviation

d is the desired precision

The following example will help us understand the concept:

> Suppose we have a process in which we want to calculate the sample size necessary to estimate the average for the weight of a tablet, with a precision of ±0.25 from its target. The historical estimate of the standard deviation is 1.0. We wish to calculate the sample size required to obtain an average within that precision (±0.25) with a 95% confidence level, that is, allowing a 5% error. Based on our analysis, the required sample size is 62. Figure 4.3 shows a spreadsheet calculation for this example.

Measurement/metric	*X* or *Y*	Type of data (discrete/continuous)	Data source and location	Sample size	Who will collect the data?	When will data be collected?	Graphical and/or statistical tools to be used

Figure 4.2 Data collection matrix.

Sample data:		
Estimate of standard deviation	S	1
Desired margin of error	delta/half-interval	0.25
Confidence level (enter .95 or 95%)	100 × (1 − α)%	95.0%
Minimum sample size	*n*	62

Figure 4.3 Sample size calculation—continuous data, example 1.

Sample data:		
Estimate of standard deviation	S	1
Desired margin of error	delta/half-interval	0.25
Confidence level (enter .95 or 95%)	100 × (1 − α)%	99.0%
Minimum sample size	*n*	107

Figure 4.4 Sample size calculation—continuous data, example 2.

Sample data:		
Estimate of standard deviation	S	1
Desired margin of error	delta/half-interval	0.10
Confidence level (enter .95 or 95%)	100 × (1 − α)%	99.0%
Minimum sample size	*n*	664

Figure 4.5 Sample size calculation—continuous data, example 3.

If we wish to increase the confidence level of our analysis to 99%, the spreadsheet tells us to increase the sample size to 107, as shown in Figure 4.4.

If we would like to calculate the sample size for a precision of ±0.10 of its target value, we simply change the 0.25 value to 0.10. Applying the new precision to our example, and keeping the 99% confidence level, our sample size increases to 664, as shown in Figure 4.5.

So far, we have calculated the sample size for continuous data. However, if the issue at hand requires the use of attribute data, the formula to be used to calculate the sample size will be

$$n = [(Z_{\alpha/2})(p)(1 - p)/(d)]^2$$

where

> Z is a constant obtained from the normal distribution table based on the allowed error
>
> p is the estimated proportion defective
>
> d is the desired precision

Let us present the concept with an example:

> Suppose the historical proportion of calls answered within the established time frame in a customer service call center is 95%. The sample size required to see a margin of error of ±3%, considering a 99% confidence level (1% error), would be 351 samples. On the other hand, the spreadsheet makes an adjustment based on the sample size and the proportion ($np \geq 5$). In this case, the spreadsheet recommends a sample size of 333 (see Figure 4.6).

It is important to realize that although the formulas provide an estimated sample size based on certain parameters, for the conclusions to be valid we need to consider the following factors:

- Data must be collected so that they are representative of the process (random).
- There must be no difference between the data collected and the data not collected.
- It is important to consider how, where, and when we sample.
- Sample from different times, so we can observe the different sources of variation of the process.
- Plot the data in a control chart to see if the process is in statistical control, which is a key requirement for many of the statistical analyses we will be performing.

Sample data:		
Estimate of proportion	P	0.95
Desired margin of error	delta/half-interval	0.03
Confidence level (enter .95 or 95%)	100 × (1 – α)%	99.0%
Minimum sample size	*n*	351
	np check (should be ≥ 5)	333

Figure 4.6 Sample size calculation—discrete data.

4.4 DESCRIBING THE SAMPLE

In Section 4.1 we defined the concept of descriptive statistics, which provide information about the data under study. When using descriptive statistics, we organize, summarize, and present the data in such a way that decisions can be taken based on them. Descriptive statistics can be categorized into three groups:

- Measures of central tendency
- Measures of dispersion
- Measures of shape

The measures of central tendency show to which point most of the data converge. These measures can be subdivided into the following values:

- Average (mean): the sum of all data points divided by the total number of data points
- Mode: the value that is repeated the greatest number of times
- Median: the value that lies in the central position once we order the data in ascending or descending order

On the other hand, measures of dispersion show how much the data vary between them. Measures of dispersion can be subdivided into the following values:

- Range: the difference between the highest and lowest value
- Variance: the square of the sum of each individual value minus the average, divided by the population size (or by sample size minus 1)
- Standard deviation: the square root of the variance

Finally, the measures of shape provide information about the type of distribution represented by the data. One of the most common graphical tools used to visualize the shape of the data is the *histogram*. This tool will be presented in more detail in the next chapter. An example of a histogram, along with descriptive statistics, for analyzing the data about the weight of a tablet (in grams) is shown in Figure 4.7.

The table shows two measures of central tendency: average (mean) and median. It also shows two measures of dispersion: range and standard deviation (stdev). Furthermore, the table provides information about the confidence intervals and the normality test. These topics will be covered later in the book.

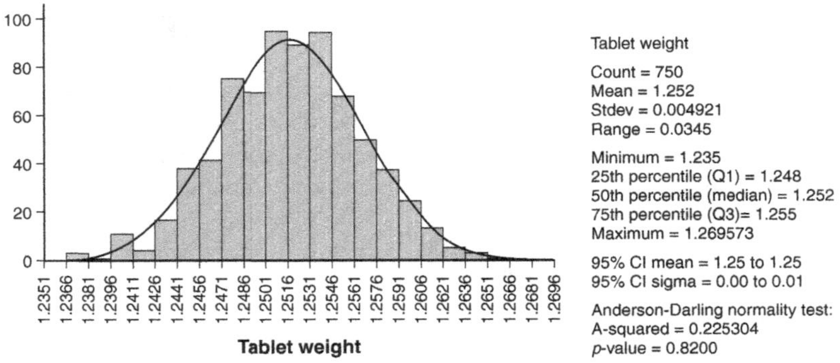

Figure 4.7 Histogram with descriptive statistics for the weight of a tablet.

4.5 THE NORMAL DISTRIBUTION

In statistics, there are many probability distributions, both for continuous data and discrete data. Among the most common probability distributions for continuous data are the normal, exponential, Weibull, lognormal, and so on. The most common probability distributions for attribute data are the Poisson, binomial, and hypergeometric. This section will cover only the normal distribution.

The *normal distribution* has certain characteristics. For instance, a normal distribution can be defined by the average and standard deviation of the population. Once we know those parameters, it can be found that 68.26% of the data will lie within ±1 standard deviation, 95.44% of the data will lie within ±2 standard deviations, and 99.73% of the data will lie within ±3 standard deviations. Later in the book we will use this concept in order to establish the statistical control limits.

Yet another characteristic of the normal distribution is that the three measures of central tendency (mode, median, and mean) are the same value. Figure 4.8 shows the relationship between the mode, median, and mean for the normal distribution.

When data do not follow a normal distribution, those three values of central tendency are not the same. Figure 4.9 shows the relationship between the mode, median, and mean for a nonnormal distribution.

What is the importance, with respect to the central tendency measures, of knowing whether the distribution is normal or nonnormal? In a normal distribution, the mode, the median, and the mean are the same value. Consequently, any of these three values represent the central tendency. However,

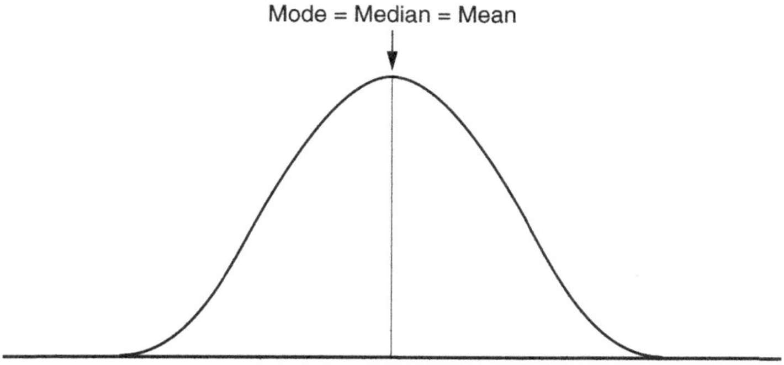

Figure 4.8 Mode, median, and mean in a normal distribution.

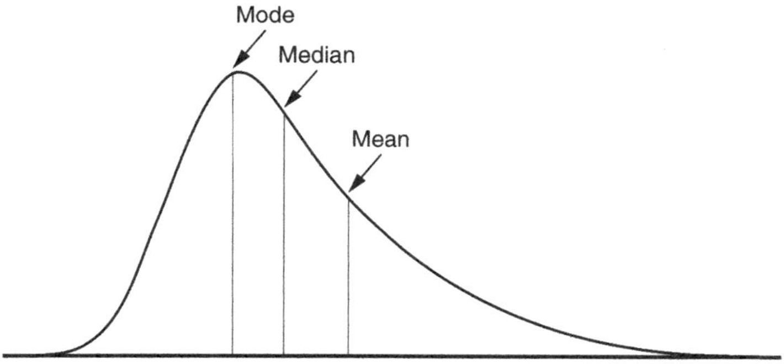

Figure 4.9 Mode, median, and mean in a nonnormal distribution.

when data do not follow a normal distribution, the median is the best estimate of central tendency because the median is less impacted by those extreme values called outliers. An *outlier* is a data point that is very small or very large when compared with the rest of the data. Because the median only considers the position of the middle datum, the median is not greatly affected by one or a few outliers. The mean, however, *is* greatly impacted by those outliers because each datum is considered in calculating the mean.

Consequently, one of the first tests to be performed when analyzing data is the *normality test*. If the data follow a normal distribution, any of the three measures of central tendency can be used. However, if data do not follow a normal distribution, the median must be used as the measure of central tendency, not the mean. Those tests that use the mean as a measure of central tendency are called *parametric* tests, while those tests that

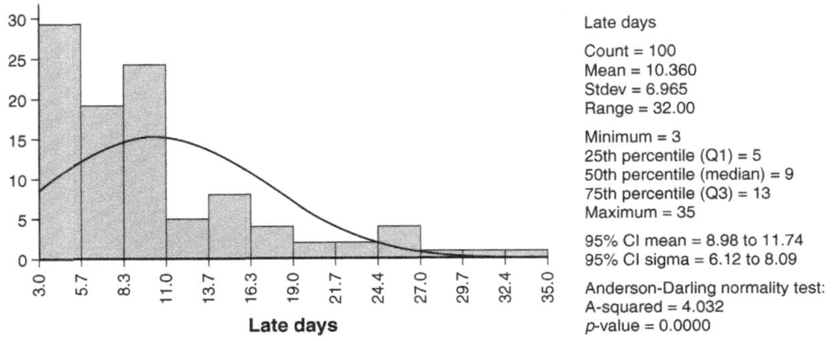

Figure 4.10 Histogram and descriptive statistics for nonnormal data.

use the median as a measure of central tendency are called *nonparametric* tests. We will be discussing these and other tests in later chapters.

It is very important to establish that although in many cases it is convenient to have data from a normal distribution, this is not always the case. Figure 4.10 is a good example.

Figure 4.10 shows the distribution of the data for late deliveries of a medical device. Ideally, the number of late deliveries would be zero. However, if there are late deliveries, we want most of them to fall within a low tardiness value. What would happen if these data followed a normal distribution? Would that be acceptable?

4.6 SUMMARY

The concept of variation is of paramount importance in order to monitor and control our processes. The difference between population and sample, parameters and statistics, discrete and variable data, and so on, must be well understood for an appropriate statistical analysis. A good sampling plan must consider factors such as type of data, process dispersion, confidence level, and precision required. Whenever possible, collect variable data instead of attribute data because the former provides more relevant information than the latter.

In order to describe the sample, we need to know about the measures of central tendency, the measures of dispersion, and the measures of shape. When data follow the normal distribution, the mode, the median, and the mean (average) are approximately the same. However, when data are skewed, the average is greatly impacted by extreme values (outliers). In this case, the median provides a better approximation of the central tendency than the average. By the same token, when data are skewed, the standard deviation and variance provide a better approximation of dispersion than the range.

In order to analyze the data, we can perform a graphical and an analytical evaluation. The following chapter presents some of the most common graphical tools used to start the evaluation of data. Just remember that graphical tools are the beginning of the analysis. The results obtained from the graphical tools have to be confirmed through the use of the analytical tools presented in subsequent chapters.

5
Graphical Tools

5.1 OVERVIEW

There are many ways to evaluate data. For instance, data can be analyzed practically. What does that mean? Evaluating the data practically means to just take a look at it and see if it makes sense. For example, is there any extreme value present? Can you see a "typo" error in the data? Any pattern? After the practical evaluation, the graphical evaluation follows. At this point in our evaluation, we rely on what the different types of graphs show. However, a graph alone does not present all the information required to make a conclusion. For instance, a histogram may show a "bell curve," but that does not necessarily mean that the data follow the normal distribution. So, in order to achieve these conclusions, an analytical evaluation of the data is required. This three-level evaluation of data can be defined as PGA: *p*ractical, *g*raphical, and *a*nalytical.

In this chapter, we will be presenting some of the most common *graphical* tools used to evaluate data. Furthermore, many practical examples of applications of graphical tools in an FDA-regulated industry will be provided throughout the chapter.

5.2 HISTOGRAM

Let us start with the histogram. This graphical tool is very helpful to analyze *continuous* data. A histogram is a bar chart that represents the frequency of certain data. Such frequency is determined by the height of each consecutive bar. Using a histogram it is very easy to graphically identify the central tendency of the data (represented by the highest bar in the graph) and the dispersion (the spread of the graph). Furthermore, using a histogram

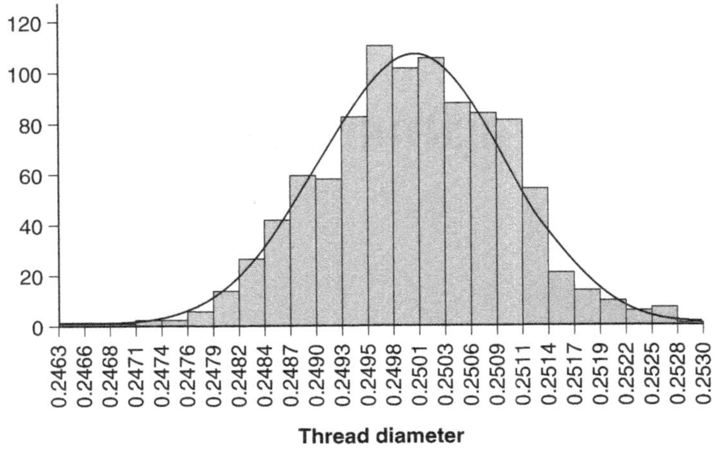

Figure 5.1 Histogram for thread diameter.

we can identify the shape of the data. Figure 5.1 shows a histogram for the diameter of a thread (in inches) used in a medical device application.

The histogram in Figure 5.1 shows the frequency for each of the dimensional intervals on a continuous scale. It also shows a normal distribution curve, which would represent that data if it were perfectly normal.

There are many times in which we do not want to analyze the totality of the data but analyze them for certain categories. An example would be segregating the data by machine. Let us suppose that data about thread diameter come from two different machines. Figure 5.2 shows the histogram for each machine in a combined chart. In this way, we can analyze the data for each machine separately.

So, whenever we want to have a quick understanding of the central tendency, dispersion, and shape of data, the histogram is an excellent graphical tool to begin our evaluation.

5.3 BOX PLOT

The *box plot*, or *box-and-whisker diagram*, is another graphical tool used to visualize the data being analyzed. The bottom of the box represents the 25th percentile, the line inside the box represents the 50th percentile (or median), and the top of the box represents the 75th percentile. The lines spreading out of the box (the whiskers) represent the expected variation.

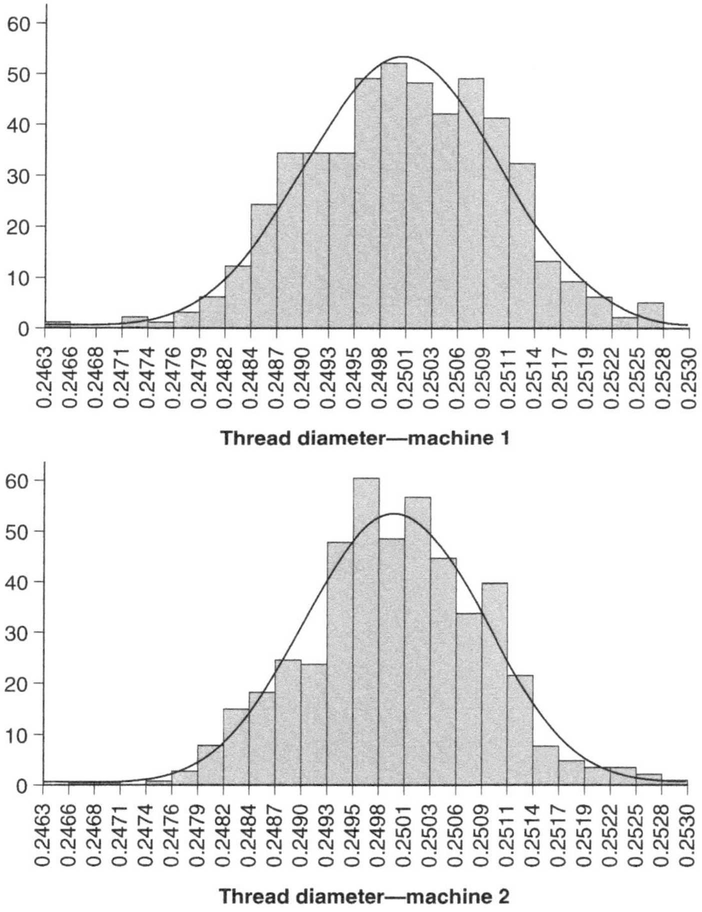

Figure 5.2 Multiple histograms for thread diameter.

Those points beyond the whiskers represent outliers. Figure 5.3 shows the box plot for our thread diameter example.

As with histograms, box plots can be developed for the totality of the data or for different categories of data. Figure 5.4 shows multiple box plots for the thread diameter example.

While histograms are mainly used to visualize the central tendency, dispersion, and shape of the data, box plots are commonly used to compare the central tendency (median or average) between certain groups. Box plots are also useful in identifying extreme values (outliers) in the data set.

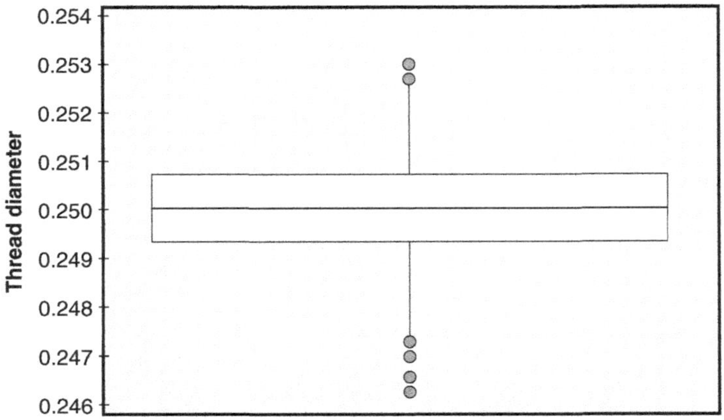

Figure 5.3 Box plot.

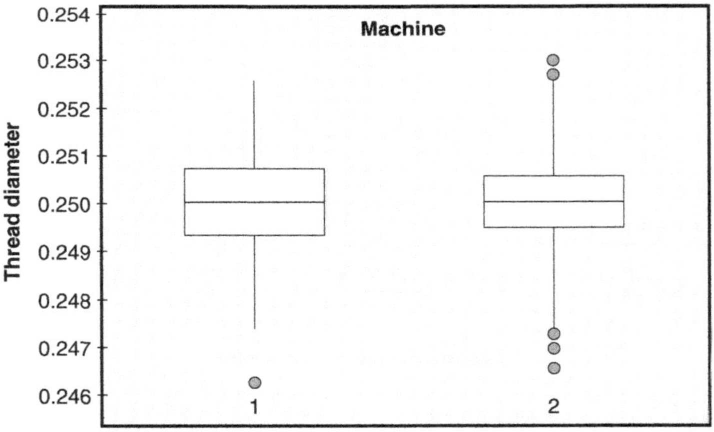

Figure 5.4 Multiple box plots.

5.4 DOT PLOT

One disadvantage of box plots and histograms is that they do not show the individual data. A graphical tool that shows each individual data point is the *dot plot*. Figure 5.5 shows the dot plot for the thread diameter data.

As with histograms and box plots, dot plots can be developed for the totality of the data or for different categories of data. Figure 5.6 shows multiple box plots for the thread diameter example.

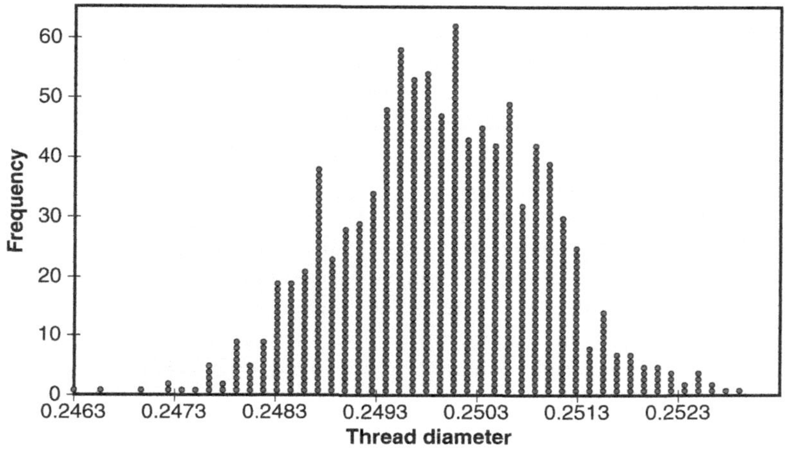

Figure 5.5 Dot plot.

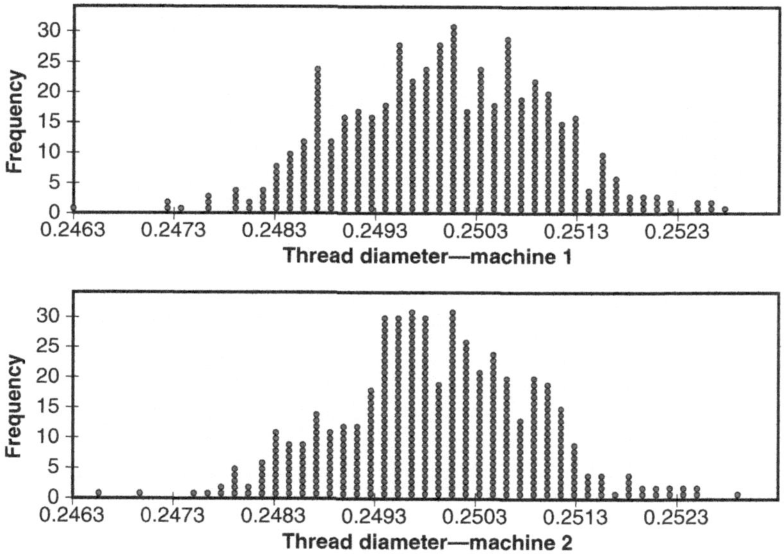

Figure 5.6 Multiple dot plots.

5.5 PARETO DIAGRAM

In Section 5.2, the histogram was presented as a graphical tool for analyzing continuous data. The *Pareto diagram*, on the other hand, is used to analyze *discrete* or *attribute* data. Specifically, the main objective of a Pareto diagram is to *prioritize*. That is the reason why the bars are represented in descending order. The most common Pareto diagram used in the quality arena is the *defects Pareto*. Used in this way, the focus of the Pareto diagram is to determine which defects have the greatest impact in our processes. Figure 5.7 shows a Pareto diagram for the packaging process in a pharmaceutical company.

The Pareto diagram shown in Figure 5.7 has two vertical axes. The axis on the left represents the frequency of each defect (represented by each bar). The axis on the right of the diagram shows the cumulative frequency when we consider each additional defect. The cumulative frequency scale is represented by the line in the upper part of the diagram.

It can be seen in the Pareto diagram in Figure 5.7 that "Cosmetic—minor scratch" is the most frequent defect, while "Incorrect lot #" is the

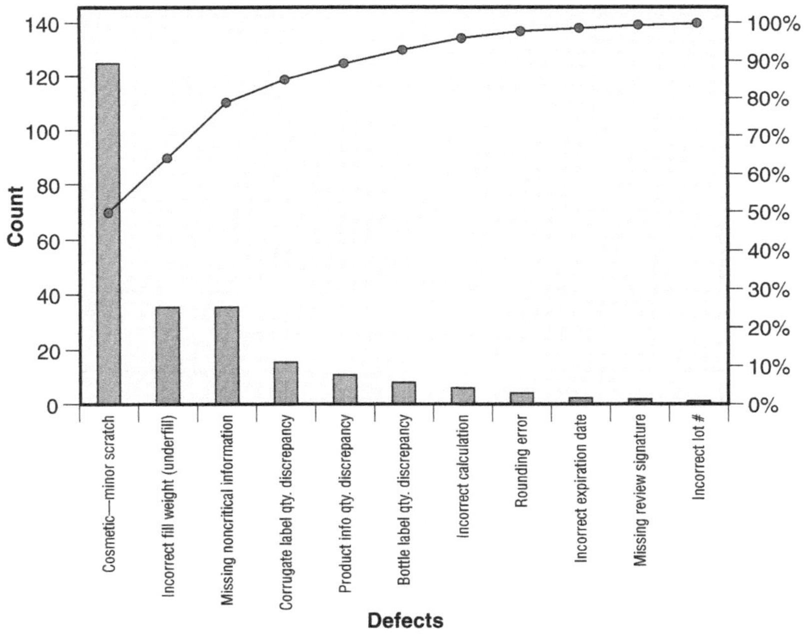

Figure 5.7 Defects Pareto diagram.

least frequent defect. Does that mean our objective would be to focus our efforts on eliminating or reducing the "Cosmetic—minor scratch" defects? Not necessarily. Why not?

Many times, the most frequent defect is not necessarily the one with the greatest impact or with the greatest cost. For this reason, it is recommended to multiply the frequency by some other factor. Some examples of that factor could be cost, severity, detectability, and so on. If the company is involved in a *failure mode and effects analysis* (FMEA) initiative, the weighting factor could be the *risk priority number* (RPN) obtained for each failure mode. More information about the practical use of the FMEA tool can be obtained from the article titled "Fail-Safe FMEA," published in the January 2012 edition of ASQ's *Quality Progress* magazine.[1] For our example, Table 5.1 shows an example of the application of a weighting factor.

Once the weighting factors are calculated, a new Pareto diagram is developed. Figure 5.8 shows the weighted Pareto diagram.

Looking at the weighted Pareto diagram, the top two defects now ("Incorrect expiration date" and "Incorrect lot #") are the two defects that appeared within the last three positions in the original Pareto diagram. Although these defects ("Incorrect expiration date" and "Incorrect lot #") occurred with the lowest frequency in the original Pareto diagram, based on their combined risk-frequency factor they are the most critical defects, which must be dealt with as the highest priority. When we look at the line

Table 5.1 Application of a weighting factor to the Pareto diagram.

Defects	Count	Criticality factor	New total
Incorect lot #	1	1000	1000
Incorrect expiration date	2	1000	2000
Cosmetic—minor scratch	125	1	125
Bottle label quantity discrepancy	9	50	450
Product info quantity discrepancy	11	50	550
Corrugate label qty. discrepancy	16	25	400
Missing noncritical information	36	10	360
Missing review signature	2	50	100
Incorrect calculation	7	25	175
Rounding error	5	50	250
Incorrect fill weight (underfill)	36	25	900

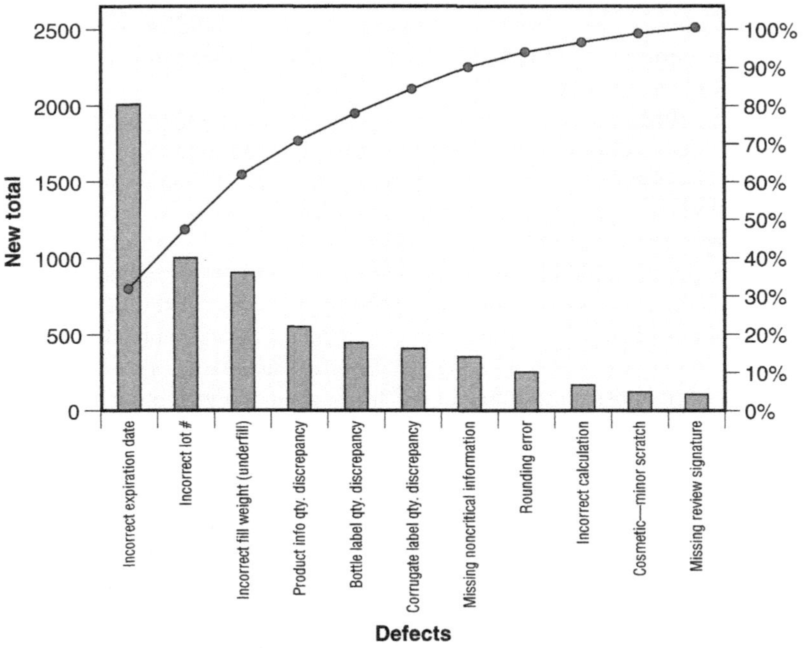

Figure 5.8 Weighted Pareto diagram.

representing the cumulative frequency, it can be noted that eliminating the top three defects will eliminate about 60% of our problems.

In the previous example, only 11 types of defects were represented in the Pareto diagram. However, as the number of defect categories increases, the Pareto diagram becomes cluttered. One important feature of the Pareto diagram is the "Other" category. In this category, those defects with the lowest impact are combined in just a single bar. In this way, we can focus our attention to those defect categories with the highest impact. Figure 5.9 shows our Pareto diagram with the "Other" bar. This bar is always presented in the last position of the diagram, regardless of its height. Remember that "Other" represents a combination of multiple types of defects.

5.6 SCATTER PLOT

Many times, we want to know if there is any kind of relationship between two variables: an input variable and an output variable. In particular, we want to know if the relationship is *positive* (as the input variable increases,

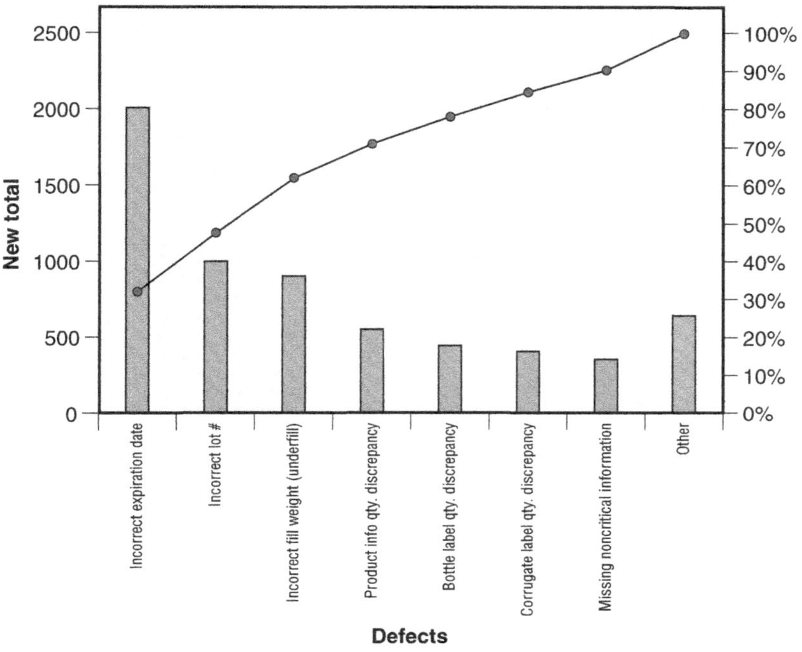

Figure 5.9 Weighted Pareto diagram with the "other" bar.

the output variable also increases) or *negative* (as the input variable increases, the output variable decreases). Also, we want to know if the relationship between those variables, regardless of whether it is positive or negative, is *strong* (dots are clustered around the regression line) or *weak* (dots are scattered on both sides of the regression line).

In order to determine if there is any relationship between two continuous variables, the *scatter plot* can be of assistance. In a scatter plot, we plot the input variable along the horizontal axis (the x-axis) and the output variable along the vertical axis (the y-axis). It is very important to note that in order to use the scatter plot, the data must be continuous for both variables. If the input variable (x) is discrete and the output variable (y) is continuous, the scatter plot would not be the most appropriate graphical tool. In this case, the box plot discussed in Section 5.3 would be more appropriate. Figure 5.10 shows shows a scatter plot in which the tablet weight (input variable) is compared to the dissolution time (output variable).

Note that the relationship between the tablet weight and the dissolution time is positive. As tablet weight increases, the dissolution time increases. There is a strong relationship between both variables, because the dots are

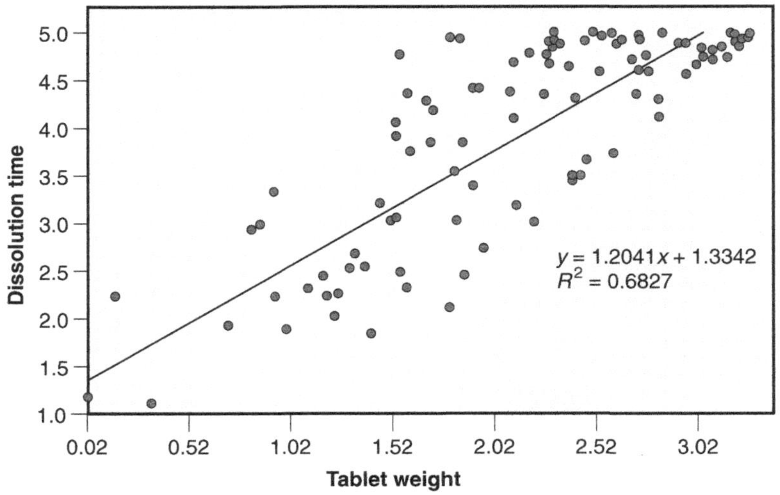

Figure 5.10 Scatter plot for tablet weight versus dissolution time.

clustered around the regression line. However, how strong is that correlation? In Chapter 9 we will discuss the concepts of *correlation coefficient* and *determination coefficient* in order to conclude how strong or weak is the relationship between the variables being analyzed.

5.7 RUN CHART

So far, the graphical tools we have seen are mostly related to gathering information and showing patterns about the central tendency, the dispersion, and the shape of the data distribution. However, none of these tools consider the order in which the data were collected (for instance, the time when the data were gathered). Take a look at the histogram presented in Figure 5.11. It represents the diameter of a pin used in a medical device. As can be noted, the data fit a normal distribution very well. When comparing the process spread with the customer specifications, the process performance index (P_p) is 1.40 and the actual process performance index (P_{pk}) is 1.34. (More information about the process capability and process performance indices will be presented in Chapter 7.)

If we only consider the shape of the data, along with the process spread compared to the customer specifications, we might conclude that the process is capable, almost centered within the specification limits, and no

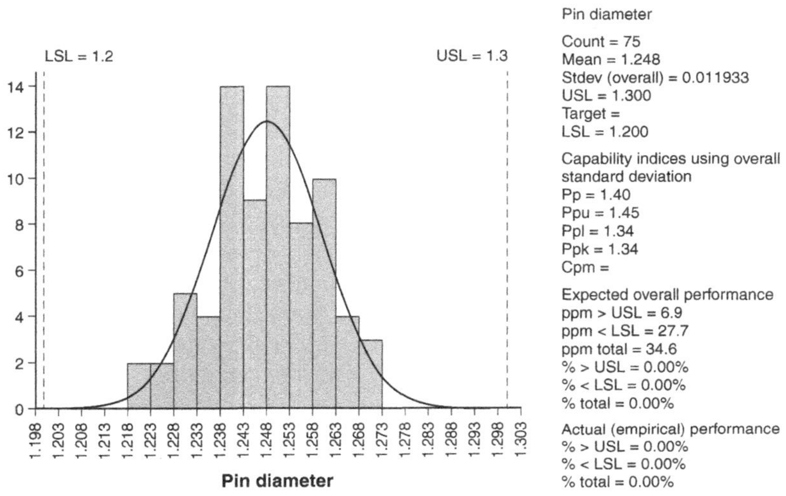

Figure 5.11 Histogram and process performance indices for pin diameter.

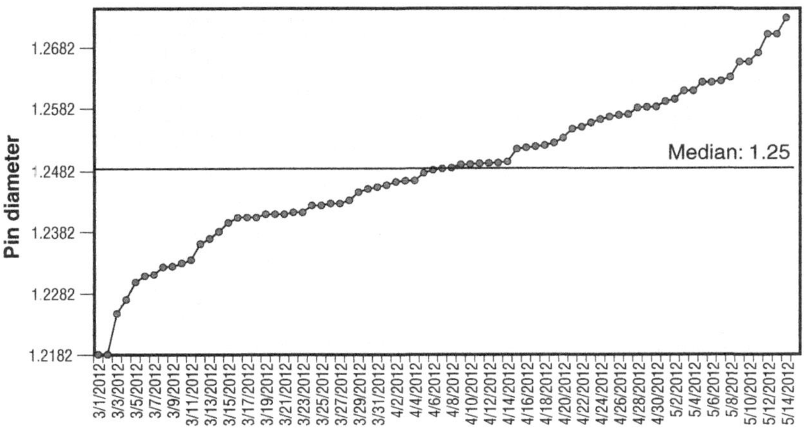

Figure 5.12 Run chart for pin diameter.

further action is required. However, when we plot the data as a run chart (that is, in the sequential order in which they were collected), as shown in Figure 5.12, we are able to see an upward trend. Since the upper specification limit was established at 1.30, it will not take long to have an out-of-specification value, probably within the next few weeks.

Organizing the data using a time-based chart is of paramount importance when data are collected in a sequential manner. Whenever we analyze any kind of data, a combination of graphical tools is preferred over just one type of graphical tool. The appropriate use of time-based charts, like the run chart and the control chart, will assist us in determining when any kind of action is needed. They can also show us when a significant change has occurred in our processes. Let us analyze the run chart in Figure 5.13.

The run chart in Figure 5.13 shows the number of days required to complete a laboratory investigation report. Each point along the horizontal axis represents an investigation (in sequential order); the value in the vertical axis represents the time it took to complete that laboratory investigation. Looking at the chart, at least three distinct patterns can be observed. The first pattern is observed for about the first 60 data points. This period represents the baseline, the period before the company started a massive CAPA system certification process. The second pattern represents the period during which the company started the CAPA system certification process. A sudden decrease in the time to complete the laboratory investigations can be observed in this period. Also, a slight reduction in the variation can be observed for the second pattern. The third pattern represents the period during which the company started to apply all the techniques learned during the CAPA system certification process. A slight reduction in the days to complete the investigation can be observed. Furthermore, a drastic decrease in the time variation can be observed during this period. This run chart can

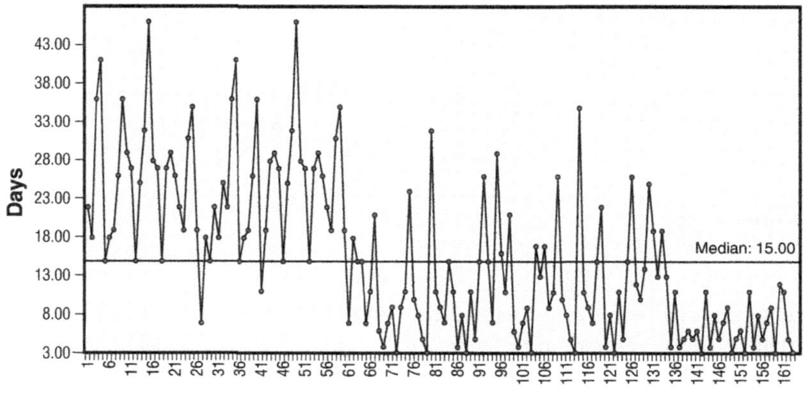

Figure 5.13 Run chart for days to complete a laboratory investigation.

be used to demonstrate the decrease in the variation and median time to complete the laboratory investigations. Furthermore, it can be used to analyze the effectiveness of the CAPA system certification training.

Most of the available statistical software packages include run charts in order to analyze different patterns. Among the patterns that most of these statistical software packages evaluate are clustering, mixtures, trends, and oscillations.

Let us start our discussion with *clustering*. Whenever a cluster is observed in data plotted in a time-based manner, that clustering might be caused by specific situations. Clusters appear as too many consecutive points on the same side of the central tendency line (the median). Possible reasons might be changes in active ingredients, different shifts, different operators, and so on. Figure 5.14 shows a run chart in which different clusters can be observed. An investigation of the cause of those clusters revealed that a change in active parmaceutical ingredient (API) occurred at several points.

The run chart is a graphical tool used to see these patterns. However, whenever we perform a run chart analysis using any statistical software package, a statistic called the *p*-value can assist us in determining if any of the previously mentioned patterns (cluster, mixture, trend, or oscillation) is being observed. For now, if we find a *p*-value lower than 0.05 for any of these possible patterns, we will conclude that the pattern is present in our data. Similarly, if the obtained *p*-value for nonrandomness is lower than 0.05, we will conclude that the data are nonrandom. Figure 5.15 shows the statistical analysis for the API example.

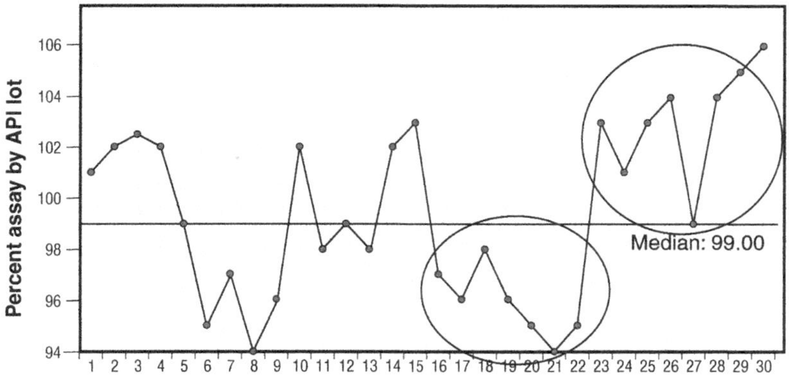

Figure 5.14 Run chart showing clusters.

Nonparametric run test: percent assay by API lot	
Number of runs about median:	9
Expected number of runs about median:	15.933
Number of points above median:	14
Number of points equal to or below median:	16
p-value for clustering:	**0.0048**
p-value for mixtures:	0.9952
p-value for lack of randomness (2-sided):	**0.0096**
Number of runs up or down:	17
Expected number of runs up or down:	19.667
p-value for trends:	0.1168
p-value for oscillation:	0.8832

Figure 5.15 Nonparametric run test showing clustering and nonrandomness of data.

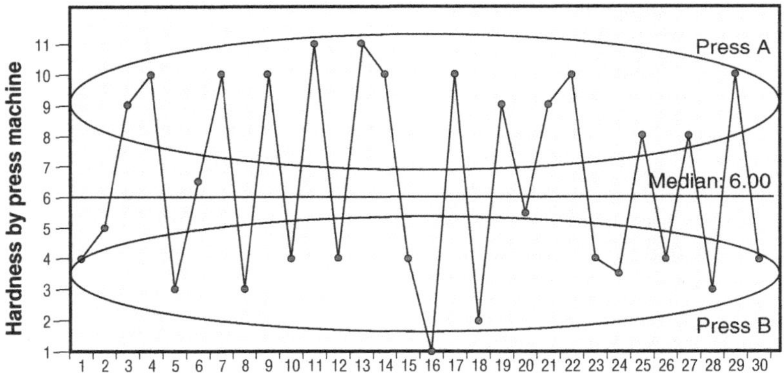

Figure 5.16 Run chart showing mixtures.

As can be seen in Figure 5.15, the data show a *p*-value for clustering lower than 0.05 (*p*-value = 0.0048), which means there is at least one cluster present. Looking at the run chart, two clusters can be observed: points 16 to 22, and points 23 to 30. As mentioned, the root cause for those shifts was the use of a different active ingredient. Looking at Figure 5.15, we can also see that the *p*-value for lack of randomness is also lower than 0.05 (*p*-value = 0.0096). That means the data are not random.

Figure 5.16 shows a run chart for data showing *mixtures*. An investigation into the cause of the mixtures revealed that hardness data came from two different machines: press A and press B.

Nonparametric run test: hardness by press machine	
Number of runs about median:	23
Expected number of runs about median:	16
Number of points above median:	15
Number of points equal to or below median:	15
p-value for clustering:	0.9954
p-value for mixtures:	**0.0046**
p-value for lack of randomness (2-sided):	**0.0093**
Number of runs up or down:	22
Expected number of runs up or down:	19.667
p-value for trends:	0.8514
p-value for oscillation:	0.1486

Figure 5.17 Nonparametric run test showing mixtures and nonrandomness of data.

As can be seen in Figure 5.17, the data show a *p*-value for mixtures lower than 0.05 (p-value = 0.0046), which means there are mixtures present. Looking at the run chart, two different populations can be observed: points above the median and points below the median. However, there are not many points close to the median. We expect most values to fall very close to the central tendency measure. However, in this case we can see a *bimodal* distribution. As mentioned, the root cause for this bimodal distribution is the use of two different press machines. In Chapter 8 we will use hypothesis tests to determine if these medians are statistically different or not. Looking at Figure 5.17, we can see that the *p*-value for lack of randomness is also lower than 0.05 (p-value = 0.0093). That means the data are not random.

Another pattern that can be analyzed with the use of run charts is a *trend*. This type of pattern can be observed whenever consecutive points show an upward or downward trend. There are countless reasons for this type of pattern. A very common root cause for this type of trend is machine wearout. Figure 5.18 shows the behavior of a pin diameter as the machine's blade starts to wear out. It can be seen that, at some point, the operator detected the pattern (around data point #11) and corrected the problem. From that point on, the pin diameter showed a random pattern.

As can be seen in Figure 5.19, the data show a *p*-value for trends lower than 0.05 (p-value = 0.0015), which means there are trends present. Looking at the run chart, an upward trend can be observed for the first 11 values. Afterward, the process started to behave in a random pattern.

Finally, the other type of pattern evaluated by most statistical software packages is *oscillation*. This type of pattern is characterized by too many consecutive jumps from one side of the central tendency line to the other

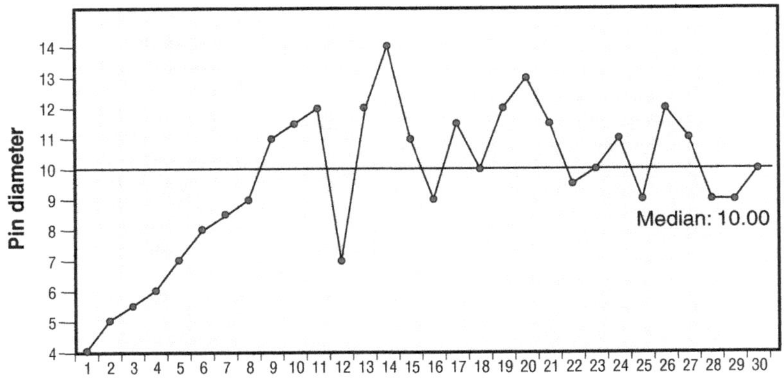

Figure 5.18 Run chart showing trends.

Nonparametric run test: pin diameter	
Number of runs about median:	13
Expected number of runs about median:	15.733
Number of points above median:	13
Number of points equal to or below median:	17
p-value for clustering:	0.1504
p-value for mixtures:	0.8496
p-value for lack of randomness (2-sided):	0.3008
Number of runs up or down:	13
Expected number of runs up or down:	19.667
p-value for trends:	**0.0015**
p-value for oscillation:	0.9985

Figure 5.19 Nonparametric run test showing trends and randomness of data.

side. This could be the result of overadjustment, which is what happened with the data shown in Figure 5.20.

A medical device manufacturer of strips was having quality problems related to strip length. An investigation revealed that the root cause of such high variation was machine overadjustment caused by a faulty sensor. Figure 5.21 shows a p-value for oscillation lower than 0.05 (p-value = 0.0086), which indicates that this pattern is being observed.

In this chapter, we have introduced the run chart as a graphical tool to identify clusters, mixtures, trends, and oscillations. These analyses will be combined with the hypothesis tests to be presented in Chapter 8.

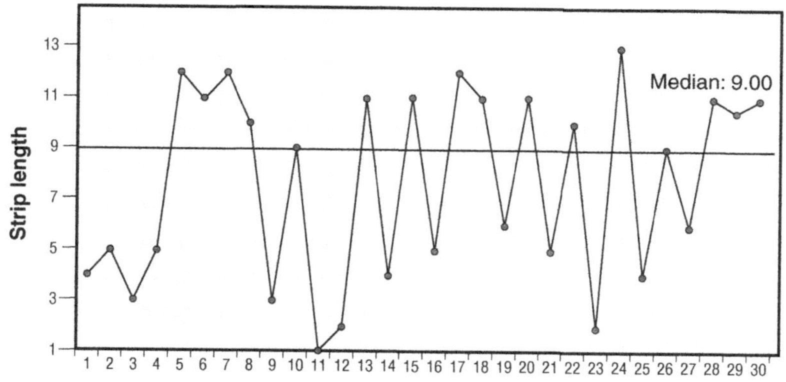

Figure 5.20 Run chart showing oscillations.

Nonparametric run test: strip length	
Number of runs about median:	16
Expected number of runs about median:	15.933
Number of points above median:	14
Number of points equal to or below median:	16
p-value for clustering:	0.5099
p-value for mixtures:	0.4901
p-value for lack of randomness (2-sided):	0.9801
Number of runs up or down:	25
Expected number of runs up or down:	19.667
p-value for trends:	0.9914
p-value for oscillation:	**0.0086**

Figure 5.21 Nonparametric run test showing oscillations and randomness of data.

5.8 NORMALITY TEST

In Section 5.2, we introduced the histogram as a graphical tool to show the shape of a data distribution. We saw how the histogram also shows the normal distribution curve to determine how well the data under study fit that normal distribution. Furthermore, using the histogram with descriptive statistics, we were able to see the estimates for central tendency, such as mean and median.

However, so far we have not examined one of the tests performed using the histogram with descriptive statistics chart. That test is called the

Anderson-Darling normality test. As will be discussed in greater detail in Chapter 8, whenever a *p*-value is lower than a probability (0.05, for illustrative purposes), a statistical hypothesis called the *null hypothesis* will be rejected. In that case, another statistical hypothesis called the *alternate hypothesis* will be accepted. Whenever a *p*-value is greater or equal to that probability, there will not be enough evidence to reject the null hypothesis.

For the Anderson-Darling normality test, the null hypothesis will be that the data follow a normal distribution. The alternate hypothesis is that the distribution is nonnormal. Figure 5.22 shows two histrograms: one of them shows a normal distribution, while the other shows a nonnormal distribution.

The upper chart in Figure 5.22 shows that the normality hypothesis can not be rejected because the *p*-value for the Anderson-Darling normality test is 0.0766 (greater than 0.05). However, the lower chart in Figure 5.22 shows a nonnormal distribution. In this case, the data are so skewed that

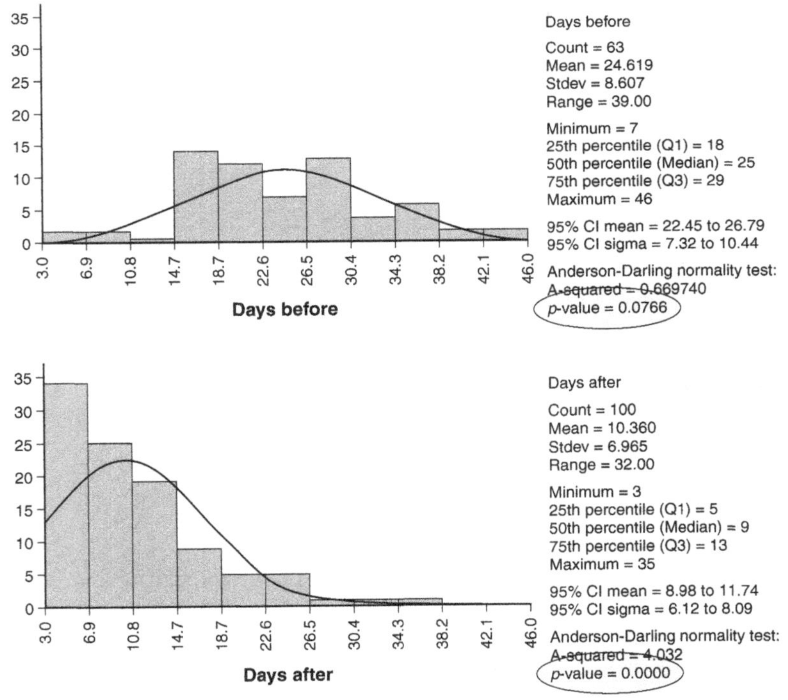

Figure 5.22 Normal and nonnormal data.

the normality hypothesis is rejected. This is evidenced by a p-value for the Anderson-Darling normality test lower than 0.05.

5.9 THE IMPORTANCE OF ASSESSING NORMALITY

So, why is it important to evaluate the normality of the data we are analyzing? In Section 4.5, the normal distribution was introduced. It was mentioned that when the data follow a normal distribution, the three measures of central tendency (mode, median, and mean) are very similar. However, when data do not follow a normal distribution (are skewed), those three measures of central tendency vary significantly. In particular, the mean (average) is the central tendency measure most sensitive to outliers.

Whenever the central tendency or dispersion of one or more distributions are going to be compared, it is important to evaluate the normality of the data. If the data follow a normal distribution, many *parametric* tests can be performed to compare means and variances. Some examples of these tests are one-sample t-test, two-sample t-test, one-way ANOVA, two-way ANOVA, F-test, and Bartlett test. On the other hand, whenever the data do not follow a normal distribution, the tests mentioned do not have any statistical value. Instead, some *nonparametric* tests could be performed on the data. Some examples of nonparametric tests are one-sample sign, one-sample Wilcoxon, two-sample Mann-Whitney, Kruskal-Wallis, and Levene tests. The details for each of these tests, along with examples applied to the FDA-regulated industry, will be presented in Chapter 8.

5.10 SUMMARY

Data can be evaluated in several ways. One common approach is to evaluate the data practically, graphically, and analytically. There are many graphical tools available. Each tool has its specific purpose. For instance, to look for central tendency, dispersion, and shape, we can use the histogram. Whenever we want to compare the central tendency of various groups, the box plot can assist us. One disadvantage of the histogram and the box plot is that they do not show the individual values; the dot plot can help us identify those individual values. If we want to prioritize the order in which we will address our quality issues, a Pareto diagram is a good option. In order to determine if there is a relationship between an independent variable (x) and a dependent variable (y), the scatter plot is an excellent tool.

Whenever data are collected in a sequential manner, they must also be presented in a way in which the time-based behavior of the data can be observed. The run chart is an easy-to-use chart to determine whether the data behave in a random manner or if some sort of pattern is being observed. Patterns such as clustering, mixtures, trends, and oscillations can be analyzed through the use of run charts. Finally, the normality of the data must be addressed because the use of certain analytical tools is subject to the assumption of that normality. If the data follow the normal distribution, certain tests (called *parametric* tests) are available, whereas if the data do not follow a normal distribution, other tests (called *nonparametric* tests) are available.

6
Measurement Systems Analysis

6.1 OVERVIEW

It was previously mentioned that every process is subject to variation. Such variation not only affects manufacturing processes, but any process. One of the least analyzed processes is the measurement process. The reason is not well known; people might not recognize that the measurement process itself is subject to variation, or we might think that because of the experience of our analysts, they are not subject to improvements. Generally, whenever we want to reduce process variation, the focus is on the manufacturing process. However, the following formula provides some insight about the components of variation:

$$\sigma^2_{\text{Total}} = \sigma^2_{\text{Process}} + \sigma^2_{\text{Measurement system}}$$

This formula establishes that total variation is equal to the process variation plus the measurement system variation. In other words, the formula urges us to consider the variation produced by the measurement system separately from the variation inherent in the processes. To summarize the point, we do not want to mask the measurement system variation with the process variation because the two types of variation are equally important.

The process used to identify the sources of variation in the measurement system is called *measurement systems analysis*; the tool used to measure those sources of variation is called *gage repeatability and reproducibility* (gage R&R). A gage R&R is a study in which several operators measure certain parts repeatedly in order to assess the *repeatability* and the *reproducibility* of the measurement system. Once the different sources of variation are analyzed, it can be determined if the variation comes from the differences between operators, differences in measurement methods, or the inherent difference between the parts. During a gage R&R, the following aspects must be considered:

- Each operator measures the same part several times.
- Data must be balanced; that is, operators must measure each part the same number of times.
- Units must represent the whole range of expected variation. It is recommended to select parts within the specification range, that is, parts from the lower specification limit, parts from the upper specification limit, and parts within the specification limits.
- Operators must measure the parts randomly. They must not know which part number they are measuring at any given time.

6.2 METRICS

Some of the metrics calculated in a gage R&R study are:

- *Repeatability*—The variation caused by the instrument. It is the variation observed when an operator measures the same part repeteadly using the same measurement instrument.
- *Reproducibility*—The variation caused by the measurement system. It is the variation observed when several operators measure the same part using the same instrument.
- *Percent precision-to-tolerance (% P/T)*—The percentage of the specification tolerance occupied by the measurement system variation.
- *Percent gage R&R (% R&R)*—The percentage of the total variation occupied by the measurement system.

For the last two metrics presented (% P/T and % R&R), a value lower than 30% is commonly desired. Specifically, a percent precision-to-tolerance (% P/T) lower than 30% helps us minimize two risks: the risk of rejecting a good part and the risk of failing to reject a defective part. The lower the % P/T, the larger the guarantee. On the other hand, a percent gage R&R (% R&R) lower than 30% allows the measurement system to observe the improvements in the process, such as the reduction in variability. When the % R&R is too high, it is possible that we could reduce variation in the process, but the measurement system (instruments, operators, and so on) might not be able to measure or observe that reduction in the variation. It is like trying to measure an improvement of a few millimeters using an instrument that only reads centimeters.

6.3 PERFORMING A GAGE R&R

An example can help us understand the theory underlying the gage R&R:

> A medical device manufacturer wants to analyze their measurement system. They produce plastic caps that will be used in a medical application. One of the most critical parameters is the inner diameter of the plastic cap. In order to evaluate their measurement system, three analysts are selected. These analysts will measure 10 parts, and each part will be measured three times by each analyst. An excerpt of the matrix generated for the study, along with the collected data, is presented in Figure 6.1.

Gage name:	Cap diameter
Date of study:	06/15/12
Performed by:	M. Peña
Notes:	

Run order	Part	Analyst	Cap diameter
1	9	1	12.26
2	7	1	10.66
3	6	2	9.80
4	8	2	10.08
5	3	1	11.34
6	6	2	10.06
7	2	1	9.32
8	8	3	9.54
9	10	3	7.84
10	5	3	8.54
11	7	3	10.21
12	2	3	8.87
13	2	1	9.42
14	10	1	8.69
15	8	1	9.83

Figure 6.1 Gage R&R data collection matrix.

Gage R&R metrics	% contribution of variance component
Gage R&R	7.76
Operator	4.37
Part × operator	0.00
Reproducibility	4.37
Repeatability	3.39
Part variation	92.24
Total variation	100.00

Figure 6.2 Percent contribution of each component.

Gage R&R metrics	% precision-to-tolerance	% Gage R&R
Gage R&R	22.68	27.86

Figure 6.3 Percent precision-to-tolerance and percent gage R&R.

Figure 6.2 shows that 92.24% of the variation is due to the parts, while the other 7.76% is due to the measurement system. Furthermore, the 7.76% variation of the measurement system can be subdivided as 4.37% due to reproducibility (differences between analysts) and 3.39% due to repeatability (the instrument variation). So, if we want to reduce the measurement system variation, we could start by analyzing what is causing the differences between analysts.

Figure 6.3 shows the % P/T and the % R&R. Note that both values are lower than 30%. These results mean that the measurement system is capable of accepting good parts and rejecting defective parts effectively, and to recognize whenever a process variation reduction is achieved.

The previous example helped us to understand the different sources of variation in the measurement system, where to focus our attention to improve the measurement system, and how reliable the measurement system is, based on the % P/T and % R&R metrics.

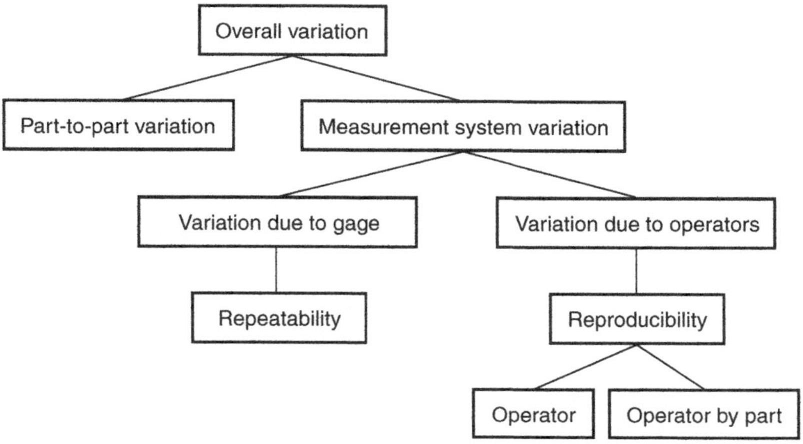

Figure 6.4 Sources of variation in a measurement systems analysis.

6.4 SUMMARY

Every system has many sources of variation. Variation can come from machines, materials, methods, measurement instruments, people, and the environment, among other sources. Very often, these sources of variation are divided into two broad categories: the variation caused by the process (part-to-part variation) and the variation caused by the measurement system. Specifically, the variation caused by the measurement system is divided into two subcategories: repeatability and reproducibility. Prior to engaging in any process variation reduction project, the sources of variability in the measurement system must be identified and reduced. Figure 6.4 summarizes the different sources of variation in our systems.

Other metrics can be calculated as well: the percent precision-to-tolerance and the percent gage R&R. These metrics are important in evaluating the reliability of our measurement system.

7
Process Capability

7.1 OVERVIEW

Once the variation due to the measurement system has been minimized, we can measure the capability of our process to produce parts within the specification limits. The tool used to compare the process variation against the customer specifications is called the *process capability analysis.*

In order to understand the concept of process capability, we need to consider two important aspects: first, the variation inherent in our process, known as *process spread*, and second, the variation allowed by the customer, known as *process specifications*. A process capability analysis combines both into a single graph, that is, how much the process varies and how much the customer allows it to vary. The process spread, also known as the *voice of the process*, is quantified by the standard deviation, σ. Particularly, the voice of the process is defined by the interval of $\pm 3\sigma$. As mentioned in section 4.5, in a normal distribution about 99.73% of the data is expected to fall within $\pm 3\sigma$ from the mean. Specifically, in Chapter 11 we will use the "$\pm 3\sigma$ *from the mean*" concept to calculate the limits for the control charts. On the other hand, the voice of the customer is defined by the process specifications, particularly, the lower specification limit (LSL) and the upper specification limit (USL). Figure 7.1 shows the process capability concept.

As long as the process spread is narrower than the process specifications, we can say that the process is *capable*, as shown in Figure 7.2. However, when the process spread is wider than the process specifications, we say the process is *incapable*, as shown in Figure 7.3.

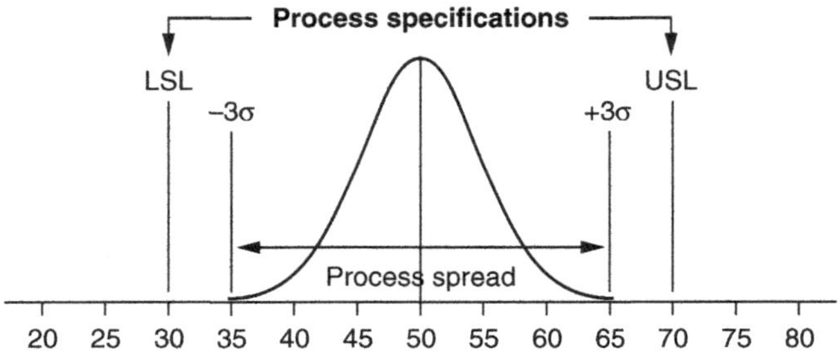

Figure 7.1 Voice of the process versus voice of the customer.

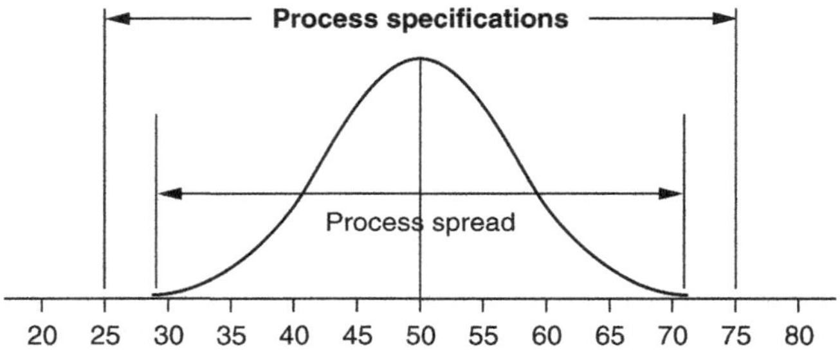

Figure 7.2 Capable process.

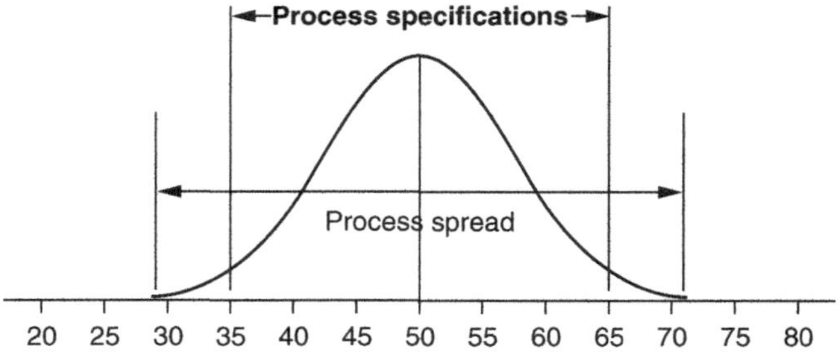

Figure 7.3 Incapable process.

7.2 PROCESS CAPABILITY AND PROCESS PERFORMANCE INDICES

In order to measure the process capability, we generally use four indices. These indices are known as C_p, C_{pk}, P_p, and P_{pk}. The former two indices, C_p and C_{pk}, are referred as *capability indices*, whereas the latter two indices, P_p and P_{pk}, are known as *performance indices*. The C_p and P_p indices are used whenever the process is centered, whereas the C_{pk} and P_{pk} indices are used whenever the process is not centered, that is, when the average is closer to one of the specification limits than the other. In this case, we will calculate two indices, C_{pk} upper and C_{pk} lower, or P_{pk} upper and P_{pk} lower, whichever applies. Then, the index with the lowest value will be selected. The rationale is that the index with the lower C_{pk} or P_{pk} shows to which side of the specification the process is shifting, that is, in which side of the specification the process is producing more defects. When the process is centered, then the probability of producing defects is the same for each side of the specification limit. Of course, that assumption only holds true as long as the process follows a normal distribution. If the data are not normal, we would need to either transform the data or use a nonnormal capability analysis. These approaches will be discussed later.

So far, we have mentioned which indices apply when the process is centered and which ones apply when the process is not centered. But, when do we use the *process capability* indices, C_p or C_{pk}, and when do we use the *process performance* indices, P_p or P_{pk}? The answer depends on whether we want to calculate the index for the long term or for the short term. The next question is, what is considered long-term and what is considered short-term? The answer does not have anything to do with how long we have been collecting the data.

What determines the short term and the long term is the way in which we calculate the process variability that goes in the denominator of the indices. Actually, there are two variability indices we will calculate: σ_R and σ_i. The σ_R value is used to calculate C_p and C_{pk}, whereas the σ_i value is used to calculate P_p and P_{pk}. The formulas for σ_R and σ_i are, respectively,

$$\sigma_R = \frac{\bar{R}}{d_2} \text{ and } \sigma_i = \sqrt{\frac{\sum (x_i - \bar{x})^2}{n-1}}$$

Figure 7.4 shows the formulas that must be used to calculate each process capability and/or process performance index, based on:

1. Whether the process is centered as compared to the specifications
2. Whether we want to calculate the short-term capability or long-term performance indices

One important issue to note is that when the process is stable, both the C_p and P_p indices should be very similar, as well as the C_{pk} and P_{pk} indices. So, the decision of which index to use must be based on the estimate of variability chosen: σ_R or σ_i. One important point must be stressed here: in order to perform a process capability analysis, the process must be in statistical control. That assumption must be met for all the statistical analyses that will be presented throughout the book. As will be seen in Chapter 11, when the process is in statistical control, there will only be common causes of variation present in the process; that is, we will have removed all the special or assignable causes from the process. An easy way to remember this point is through the following adage: "*There is no capability without stability.*" So, prior to performing the process capability analysis and drawing conclu-

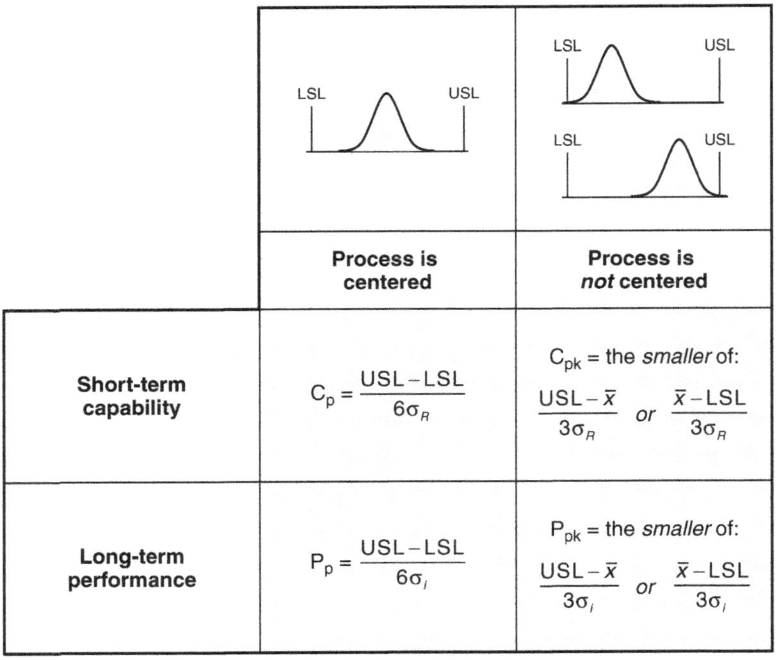

Figure 7.4 Process capability and process performance indices.

sions from the calculated indices, make certain that the process is stable (in statistical control). This assumption can be verified through the use of a control chart. Most statistical software includes control charts as part of the process capability analysis. Remember, the next time that you see a process capability analysis, ask for the control charts and make certain that the process is in control. If the process is not in control, the calculated indices might not have any statistical meaning.

7.3 HOW TO INTERPRET THE PROCESS CAPABILITY AND PROCESS PERFORMANCE INDICES

Many times, we hear about the misuse of process capability indices. Some companies use the C_{pk} index to determine when the process is capable, while other companies use the P_{pk} index to achieve the same goal. Neither C_{pk} nor P_{pk} can be used *alone* in order to determine how capable the process is. Each one (C_{pk} or P_{pk}) must be used *in combination* with the C_p or P_p in order to address the overall process capability. First, it must be understood that C_p and P_p only take into consideration the process specification versus the process spread, as shown in Figure 7.4. It does not consider the process centering. So, C_p and P_p will be used to determine process capability, regardless of process centering. On the other hand, C_{pk} and P_{pk} will be used to determine process centering, regardless of process capability. Figure 7.5 provides an example of different scenarios that we might encounter and how each scenario will be interpreted. This is very important to know because each scenario will require a different approach to improve the process. For this example, we will only compare C_p and C_{pk}. However, the same approach can be used for P_p and P_{pk}. Remember that the difference

$C_p = 0.75$	$C_{pk} = 0.75$	Not capable Centered
$C_p = 1.33$	$C_{pk} = 0.95$	Capable Not centered
$C_p = 0.95$	$C_{pk} = 0.75$	Not capable Not centered
$C_p = 1.33$	$C_{pk} = 1.33$	Capable and Centered

Figure 7.5 Interpretation of process capability and process performance indices.

between C_p (or C_{pk}) and P_p (or P_{pk}) is the way in which we calculate the process variability (σ_R or σ_i).

For instance, in the scenario where $C_p = 0.75$ and $C_{pk} = 0.75$, we can conclude that the process is not capable (since $C_p < 1.0$). However, since both C_p and C_{pk} are the same, we can conclude that the process is centered. In this case, our problem is about process variation. If we want to increase both C_p and C_{pk}, either the process specifications must be widened (less realistic) or the process spread must be narrowed (more realistic). On the other hand, in the scenario where $C_p = 1.33$ and $C_{pk} = 1.33$, we can conclude that the process is capable and centered. This is the ideal scenario.

Furthermore, in the scenario where $C_p = 1.33$ and $C_{pk} = 0.95$, we can conclude that the process is capable. However, since C_p and C_{pk} are not the same, we can conclude that the process is not centered. In this case, our problem is about process centering, not about process variation. If we want to increase the C_{pk}, we must center the process. One misconception lies in concluding that the process is incapable by having a C_{pk} index lower than 1.0. As mentioned above, if the C_p is high (for example, $C_p = 1.33$) and the C_{pk} is low (for example, $C_{pk} = 0.95$), the process is still capable because the process spread is narrower than the process specifications. In this case, the process is fixed by just adjusting the centering. Here is an analogy to help remember this concept: as long as your car is narrower than your house's garage, you are capable to park the car inside the garage, regardless of whether you center it or not. Not centering the car does not make the parking process incapable. The only fact that would make that process capable is if your car is wider than the garage. In that case, even by centering the car in the garage, you will not be able to park the car inside the garage.

Finally, the worst scenario is presented where $C_p = 0.95$ and $C_{pk} = 0.75$. In this scenario the process is neither capable nor centered. If we want to improve our process, we must first center it. In this way, both C_p and C_{pk} can be increased to 0.95 without any further changes. However, in order to increase the C_p and C_{pk} above the acceptable 1.33 value, we must either widen the process specifications (less realistic) or the process spread must be narrowed (more realistic).

7.4 PROCESS CAPABILITY ANALYSIS FOR NONNORMAL DATA

Very often, organizations deal with processes that produce nonnormal data. So far, our analyses have focused on dealing with normal data. But what can we do when we obtain nonnormal data from our processes? The two most common approaches are to transform the data (for example, using a Box-Cox transformation or Johnson transformation) or to obtain a probability distribution for which the data make a fit. Out of those two approaches, we are going to develop an example using the Box-Cox transformation:

> Let us suppose that a pharmaceutical company is analyzing the assay percent parameter for a certain product. The first graphical tool they used to analyze the data was a histogram and descriptive statistics summary, as shown in Figure 7.6. From the Anderson-Darling normality test of such analysis, we can conclude that the data do not follow a normal distribution. This conclusion is based on the obtained p-value for the test of 0.0478. As mentioned earlier, using a 95% confidence level, when the obtained p-value is lower than 0.05, we reject the hypothesis that the data follow a normal distribution. So, normality of the data is discarded. Figure 7.6 shows that the data are positively skewed, or skewed to the right.
>
> If we perform a process capability analysis, erroneously assuming the data follow a normal distribution, we would obtain the results shown in Figure 7.7. The P_p index for such data would be 1.31, and the P_{pk} index for the data would be 1.30. Both values do not differ much from the accepted value of 1.33. Furthermore, we can say the process is centered between the specification limits because the P_p and the P_{pk} values are approximately the same. In terms of defective parts produced, the analysis shows approximately 90 parts per million defective.
>
> In fact, based on the results, we would conclude that the process is doing fairly well, and with a small reduction in the variation we could obtain P_p and P_{pk} values above 1.33. However, the

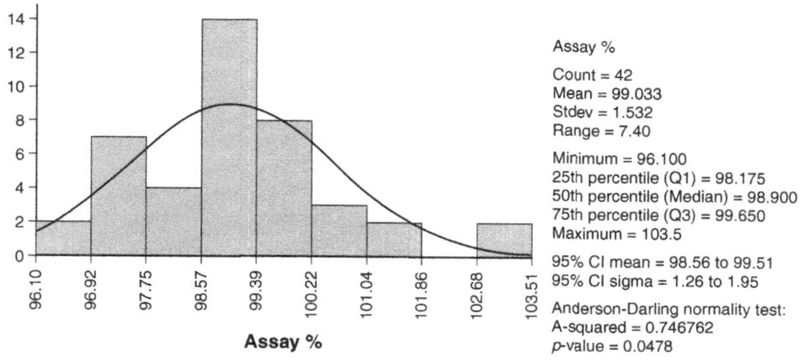

Figure 7.6 Histogram and descriptive statistics for nonnormal data example.

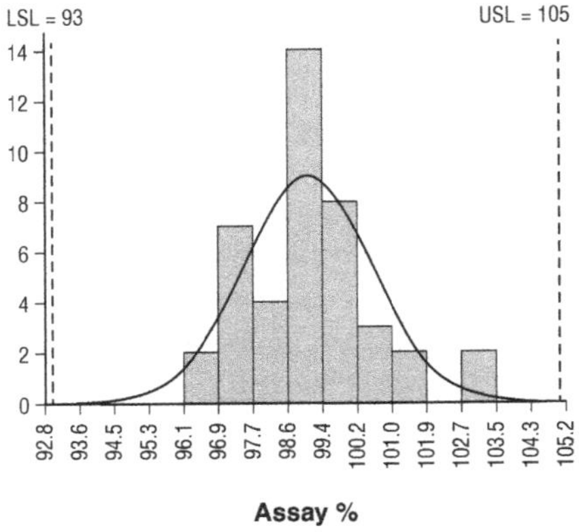

Process capability report: Assay %	
Count	42
Mean	99.033
StDev (overall, long term)	1.532
StDev (within, short term)	1.399
USL	105
Target	
LSL	93

Performance indices using overall StDev	
Pp	1.31
Ppu	1.30
Ppl	1.31
Ppk	1.30
Cpm	

Expected overall performance	
ppm > USL	49.3
ppm < LSL	41.1
ppm total	90.401
% > USL	0.00%
% < LSL	0.00%
% total	0.01%

Figure 7.7 Normal process capability analysis for nonnormal data example.

conclusion will be wrong because we assumed a normal distribution for the data when, in fact, the data are skewed to the right. So, in order to obtain a more accurate result about the P_p, P_{pk}, and defective parts per million, we will perform an analysis using a Box-Cox transformation.

Transforming data means performing the same mathematical operation on each piece of original data. The statisticians George Box and David Cox developed a procedure to identify an appropriate exponent (lambda, or λ-value) to use to transform data into a "normal shape." The λ-value indicates the power to which all data should be raised. In order to do this, the Box-Cox power transformation searches from $\lambda = -5$ to $\lambda = +5$ until the best value is found. Using the Box-Cox power transformation in a statistical analysis software program provides an output that indicates the best λ-values.

For our example, a Box-Cox transformation was performed using statistical software, obtaining an optimal λ-value = –5. This means that all data values (Y) will be raised to a power of –5, or a transformed value equal to $Z = Y^{-5}$. The same transformation we performed on the individual values has to be performed on the specification limits, USL and LSL. A standard deviation will be calculated from the transformed data, and the C_p, C_{pk}, P_p, or P_{pk} indices can be calculated as shown earlier in Figure 7.4. Doing such

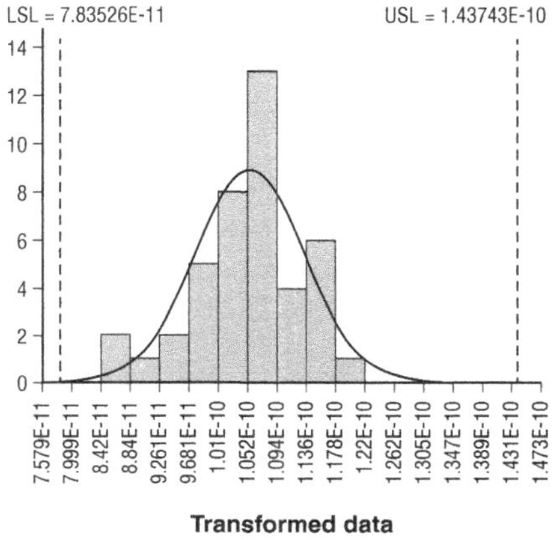

Process capability report (nonnormal): Assay %	
Box-Cox transformation: Z = Y ^ (–5.0000)	
Sample count	42
Sample mean	99.033
USL	105
Target	
LSL	93
Lambda (optimal)	–5
Mean (transformed)	0.00
StDev (transformed overall, long term)	7.88718E-12
StDev (Transformed within, short term)	7.2711E-12
USL (transformed)	7.83526E-11
Target (transformed)	
LSL (transformed)	1.43743E-10

Performance indices using transformed overall StDev	
Pp	1.38
Ppu	1.14
Ppl	1.62
Ppk	1.14

Expected overall performance	
ppm > USL	311.8
ppm < LSL	0.6
ppm total	312.4
% > USL	0.03%
% < LSL	0.00%
% total	0.03%

Figure 7.8 Box-Cox transformation process capability analysis for nonnormal data example.

calculations for the P_p and P_{pk} indices, we obtain $P_p = 1.38$ and $P_{pk} = 1.14$, as shown in Figure 7.8.

The values obtained in Figure 7.8 differ from the $P_p = 1.31$ and $P_{pk} = 1.30$ obtained in Figure 7.7, when we assumed the data followed a normal distribution. Furthermore, for the analysis assuming normal data, we obtained approximately 90 defective parts per million. However, based on the analysis with the Box-Cox transformation, approximately 312 defective parts per million are produced in this process. This means more than three times the expected defective parts per million when compared with the first analysis.

Although the Box-Cox power transformation is frequently used to transform nonnormal data, it is not a guarantee for normality. This is because it actually does not really check for normality. The method checks for the smallest standard deviation. The assumption is that among all transformations with λ-values between –5 and +5, transformed data has the highest likelihood (but not a guarantee) to be normally distributed when the standard deviation is the smallest. Therefore, it is absolutely necessary to always check the transformed data for normality using a probability plot.

Anderson-Darling normality tests	
AD normality transformed data	0.541699
AD normality *p*-value transformed data	0.1549
AD normality original data	0.746762
AD normality *p*-value original data	**0.0478**

Figure 7.9 Normality test for original data and Box-Cox transformed data.

In our example, the Anderson-Darling normality test's p-value for the original data was 0.0478, and for the Box-Cox transformed data is 0.1549, as presented in Figure 7.9. Based on this, we can not reject the hypothesis that the transformed data follow a normal distribution.

7.5 PERFORMING A PROCESS CAPABILITY ANALYSIS

Let us now perform a complete process capability analysis to evaluate the current status of a process and evaluate the status of such process after an improvement project has been completed:

> A cough syrup manufacturer is facing some quality problems in their filling process. During the sealing process, some overfilled bottles are not sealing completely. In order to evaluate the current status for the *net weight* parameter, a histogram and descriptive statistics summary was developed. The summary is presented in Figure 7.10.
>
> An individuals and moving range (ImR) chart was developed in order to determine whether the process is stable, that is, if only common causes of variation are present. Remember that prior to performing a process capability analysis, you need to make certain that the process is in statistical control. Or, said in another way, "There is no capability without stability," as you remember. The resulting individuals and moving range chart is presented in Figure 7.11.
>
> Now that we know that only common causes of variation are present in our process, we can proceed with the capability analysis. Since the Anderson-Darling normality test in Figure 7.10 shows a p-value greater than 0.05, a normal capability analysis

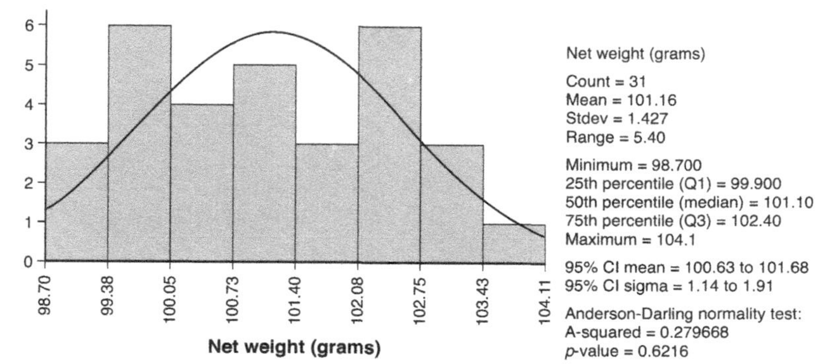

Figure 7.10 Histogram and descriptive statisistics for net weight.

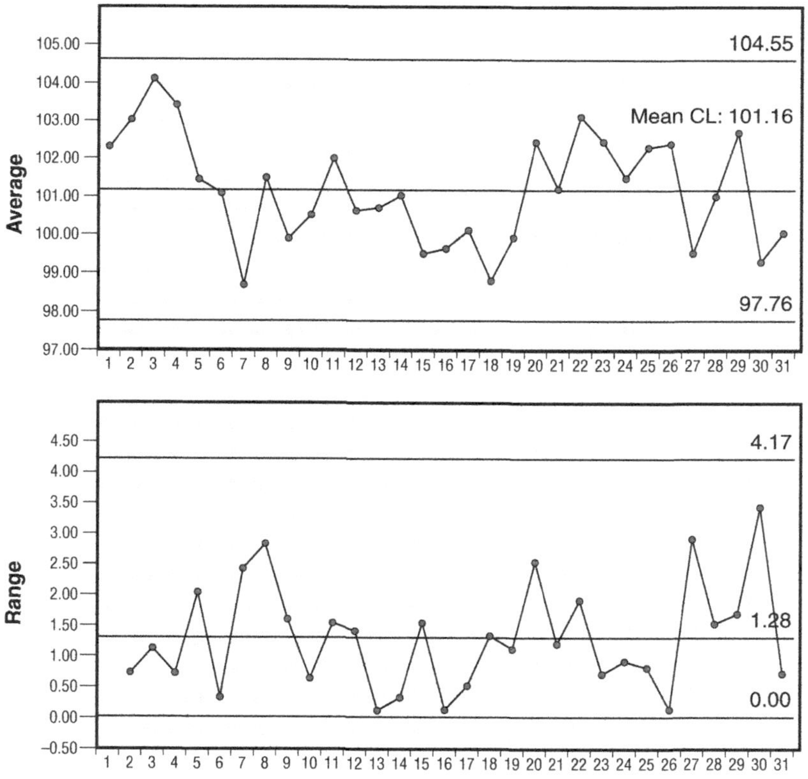

Figure 7.11 Individuals and moving range chart for net weight.

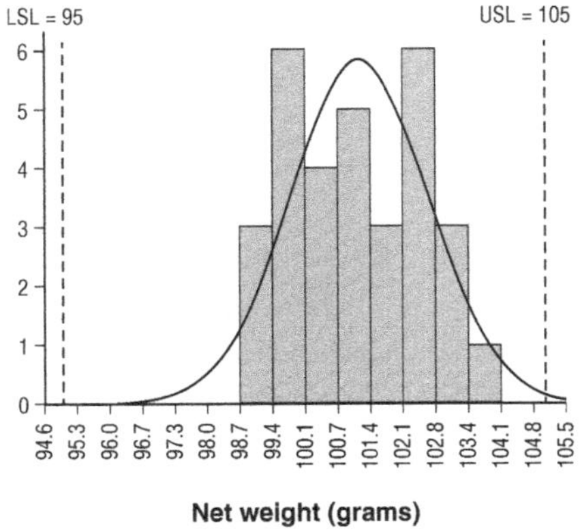

Process capability report: Net weight (grams)	
Count	31
Mean	101.16
StDev (overall, long term)	1.427
StDev (within, short term)	1.132
USL	105
Target	
LSL	95

Capability indices using within StDev	
Cp	1.47
Cpu	1.13
Cpl	1.81
Cpk	1.13
Cpm	

Expected overall performance	
ppm > USL	3544.5
ppm < LSL	7.9
ppm total	3552.5
% > USL	0.35%
% < LSL	0.00%
% total	0.36%

Figure 7.12 Normal process capability analysis for net weight.

will be performed for the data. Figure 7.12 shows the capability analysis results for the net weight parameter.

As you can observe in Figure 7.12, the process is capable (C_p = 1.47); however, it is not centered between the specifications (C_{pk} = 1.13). Furthermore, it can be observed that the average is shifted to the right of the target value of 100 grams. It is such a shift toward the upper specification limit that is causing the overfilling problem. During a brainstorming session, it was identified that the process was intentionally set toward the upper specification limit because some complaints about underfilled bottles were received during the previous year.

So, the improvement team is faced with two opportunity areas: (1) eliminating the problem caused by improperly sealed bottles, while (2) eliminating any possibility of a complaint due to underfilled bottles. After various design of experiments (to be discussed in Chapter 10) were performed, the proper machine settings were established in order to achieve the target of 100 grams while reducing the variation within the filling process. The results for the next 50 bottles after the improvement, along with the 31 bottles before the improvement, are shown in Figure 7.13.

It can be seen in Figure 7.13 that a decrease in the net weight was achieved from the improvement project, as well as a major

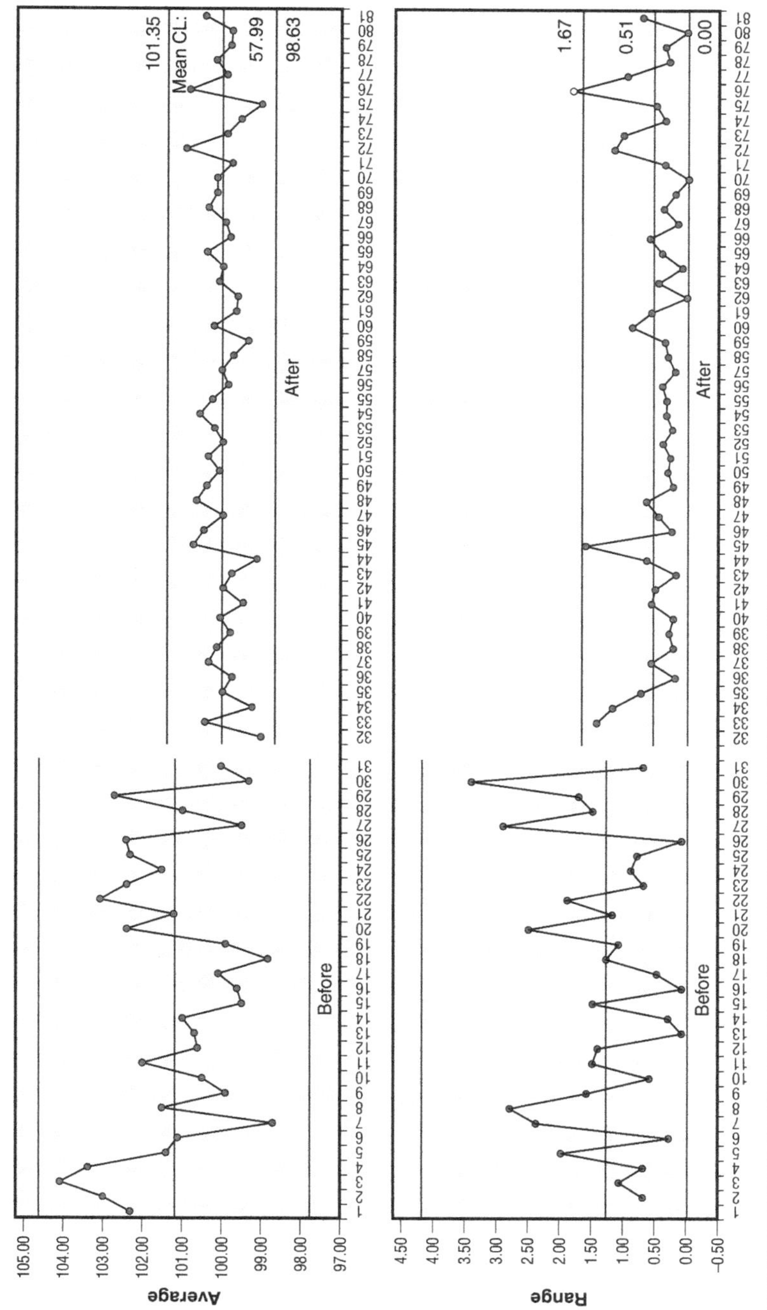

Figure 7.13 ImR chart for before and after analysis for net weight.

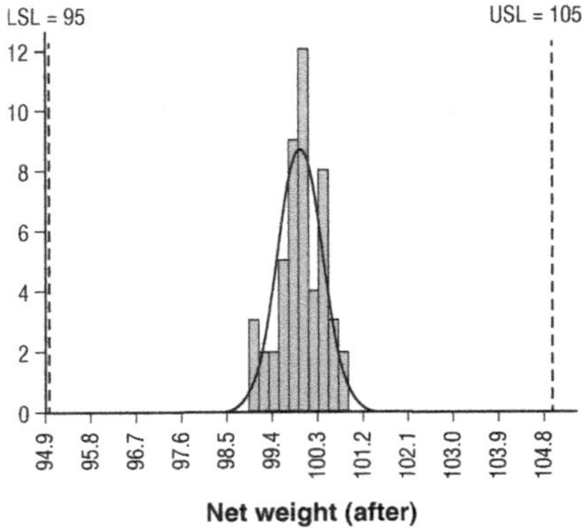

Process capability report: Net weight (after)	
Count	50
Mean	99.989
StDev (overall, long term)	0.442037
StDev (within, short term)	0.453685
USL	105
Target	
LSL	95

Potential capability indices using within StDev	
Cp	3.67
Cpu	3.68
Cpl	3.67
Cpk	3.67
Cpm	

Expected overall performance	
ppm > USL	0.0
ppm < LSL	0.0
ppm total	0
% > USL	0.00%
% < LSL	0.00%
% total	0.00%

Figure 7.14 Process capability analysis for net weight after the improvement project.

decrease in the variability of the filling process. In order to determine the effect of such improvements on the C_p and C_{pk} indices, a process capability analysis was performed, as shown in Figure 7.14. After the improvent project, the C_p index went from 1.47 to 3.67, while the C_{pk} index went from 1.13 to 3.67. Now the process is capable and meeting its target value. Furthermore, the defective parts per million were reduced from approximately 3552 bottles to zero, as shown in Figure 7.14.

Now that we know how to evaluate and optimize our measurement systems and how to determine if our processes are capable, we can proceed to make comparisons between population means, medians, variances, and/or proportions through the hypothesis tests.

7.6 SUMMARY

The process capability analysis is used to compare the process spread against the process specifications and evaluate how capable the process is for producing within those specification limits. Prior to performing a process capability analysis, we need to make certain the process is in statistical control. In order to verify that such assumption is met, a control chart can be used. Remember that "there is no capability without stability." Once we know the process is in statistical control, we need to evaluate normality of the data to determine if we are going to use a normal capability analysis or a nonnormal capability analysis (or maybe use a transformation of the data). Then, we need to decide which indices to calculate: C_p and C_{pk}, or P_p and P_{pk}. Just remember that the main difference in the calculation of these indices is which estimate of the variation to use: σ_R or σ_i. The σ_R is used to calculate C_p and C_{pk}, while the σ_i is used to calculate P_p and P_{pk}. Once the indices are calculated, we can translate that result into defective parts per million produced by the process. The obtained results can then be used as a baseline for our process variation reduction projects. Let us start the reasoning for potential projects by making comparisons between different groups, using a tool called hypothesis testing.

8
Hypothesis Testing

8.1 OVERVIEW

A *hypothesis test* is a method of making decisions using data from our processes. In statistics, a result is called *statistically significant* if it is unlikely to have occurred by common causes of variation alone, according to a predetermined threshold probability called the *significance level*. Hypothesis tests answer the question "Assuming that the null hypothesis is true, what is the probability of observing a value for the test statistic that is at least as extreme as the value that was actually observed?" That probability is known as the *p*-value. In simple terms, the *p*-value might be interpreted as the confidence we have in the null hypothesis.

If the *p*-value is less than the required significance level, then we say the null hypothesis is rejected at the given level of significance. Rejection of the null hypothesis is a conclusion. This is like a "guilty" verdict in a criminal trial (the evidence is sufficient to reject innocence, thus proving guilt). Under this scenario, we might accept the alternate hypothesis. If the *p*-value is not less than the required significance level, then the test has no result. The evidence is insufficient to support a conclusion. This is like a jury that fails to reach a verdict, because guilt was not proven beyond a reasonable doubt. To summarize, here is a slogan you can use to remember when to reject or fail to reject the null hypothesis (H_0): "If *p*-value is *low*, the null hypothesis must *go*." In other words, when the *p*-value for the null hypothesis is lower than a predetermined value, then the null hypothesis must be rejected.

Whether rejection of the null hypothesis truly justifies acceptance of the alternate hypothesis depends on the structure of the hypotheses. Rejecting the hypothesis that a large bite originated from a tiger does not immediately prove the existence of a gargoyle. Hypothesis testing emphasizes the *rejection*, which is based on a probability, rather than the *acceptance*, which requires extra steps of logic. When dealing with hypothesis testing, there are four decisions that can be made. Two of those decisions are correct decisions, while two of them are wrong decisions. In statistics, those wrong decisions are called *type I error* and *type II error.* Figure 8.1 shows the concept of these four possible decisions, using an example about the acceptance or rejection decision to be made on a production lot:

> In any manufacturing process, the assumption must be that all lots produced comply with the quality requirements; that is, our null hypothesis is that the lots are good. Based on that assumption, the alternate hypothesis would be the opposite; that is, the lots are defective. Whenever an inspector is faced with a lot (good or defective) he or she can make one of two decisions: either to reject the lot or to accept the lot (which in hypothesis testing terms would be "fail to reject the lot").
>
> Taking a look at Figure 8.1, let us first explore the two good decisions. If an inspector receives a good lot and accepts it (fail to reject), that would be a good decision. The probability of failing to reject a good lot can be calculated as $1 - \alpha$, which we expect to be a very high probability. In this instance, the good decision will be called the *producer's confidence* because such decision will be very beneficial to the producer of that lot. The opposite (reject-

	Null hypothesis: Good lot	**Alternate hypothesis: Defective lot**
Decision: Fail to reject the lot	Good decision $1 - \alpha$ Producer's confidence	Type II error Probability = β Consumer's risk
Decision: Reject the lot	Type I error Probability = α Producer's risk	Good decision $1 - \beta$ Consumer's confidence

Figure 8.1 Possible decisions in the acceptance or rejection of a lot.

ing a good lot) would be called the *producer's risk*, as I will discuss later. The second good decision in our example would be to reject a defective lot. The probability of rejecting a defective lot can be calculated as $1 - \beta$, which we also expect to be a very high probability. In this instance, the good decision will be called the *consumer's confidence* because such decision will be very beneficial to the consumer of that lot. The opposite (failing to reject a defective lot) would be called the *consumer's risk*, as will be discussed later.

Let us take a look now at the two wrong decisions, namely type I error and type II error. Whenever an inspector rejects a good lot, he or she is making a type I error. The probability of that error is established by a value called α, which we expect to be a very low probability. In this instance, the wrong decision will be called the *producer's risk* because such decision will be very detrimental to the producer of that lot. The second wrong decision in our example would be to fail to reject a defective lot. Such probability is established as β, which we also expect to be a very low probability. In this instance, the wrong decision will be called the *consumer's risk* because such decision will be very detrimental to the consumer of that lot.

So, the obvious question is "Which of these errors is more important to avoid?" The answer is that both errors are important. However, guarding against one type of error could result in an increase in the other type of error. The best strategy to reduce both errors is to increase the sample size to perform the appropriate hypothesis test. Doing so will allow us to detect smaller shifts due to nonrandom causes of variation.

Prior to performing any hypothesis test, there are some assumptions that must be satisfied. For instance, when dealing with variable or continuous data, the assumption is that the data follow a normal distribution. If data are not normal, you would probably have to perform a transformation of the data, as previously mentioned.

When we are comparing groups from different *populations*, the following assumptions must be satisfied:

- Samples are independently collected.
- Samples are obtained randomly.
- Samples are representative of the population.

When we are comparing groups from different *processes*, the following assumptions must be satisfied:

- Each process is stable.
- There are no special causes or shifts over time.
- Samples are representative of the process.

Now that we know the basics about hypothesis testing, let us explore some tests to compare means, medians, and variances. In Section 5.9 we discussed the importance of assessing the normality of the data in order to determine which type of tests we have available to compare groups. When the data follow a normal distribution, the set of tests to compare groups is called *parametric* testing. When the data do not follow a normal distribution, the set of tests is called *nonparametric* testing. Let us start with the most common parametric tests, those used to compare means.

8.2 COMPARING MEANS

The four most common tests to compare means when the data follow a normal distribution are one-sample *t*-test, two-sample *t*-test, one-way ANOVA test, and two-way ANOVA test. The choice about which test must be used has to do mainly with how many means we want to compare: one mean against a fixed value, one group's mean against another group's mean, or the means among three or more groups.

8.2.1 One-Sample *t*-Test

The simplest test is when we compare one group's mean against a fixed value. That fixed value could be a specification, a standard, a target value, an improvement value, and so on. Let us illustrate the application of the one-sample *t*-test with an example:

> Three months ago, a company implemented a new CAPA investigation process after a massive training. The company wants to know if the training is paying off, that is, if the average closure time for the investigations has decreased significantly. Before training, average closure time was 20 days. Figure 8.2 shows the results of the one-sample *t*-test using statistical software.
>
> It can be noted that the mean cycle time of the 100 samples taken after the training's completion was reduced from 20 days to 10.36 days. The question: "Is the observed difference the result of the training or just the result of common causes of variation?" The null hypothesis was established as a mean cycle time equal to 20 days, while the alternate hypothesis was established as a mean

One-sample *t*-test	
Test information	
H_0: Mean (Mu) = 20	
H_a: Mean (Mu) less than 20	
Results	**Cycle time (days)**
Count	100
Mean	10.360
StDev	6.965
SE Mean	0.696531
t	−13.840
p-value (1-sided)	**0.0000**
UC (1-sided, 95%)	11.517

Figure 8.2 One-sample *t*-test example.

cycle time of less than 20 days, which is what we would expect. Recall from Section 8.1 that the criterion to determine if the null hypothesis will be rejected or not is going to be the *p*-value. In that section I mentioned that "*if* p-*value is low,* H_0 *must go.*" In our examples, for consistency purposes, I will use a producer's confidence of 95%, which means that $\alpha = 0.05$, or a significance level of 5%.

Taking a look at Figure 8.2, the obtained *p*-value = 0.0000 (not necessarily zero, but a very small number). We already mentioned that if the *p*-value is less than the required significance level, then we say the null hypothesis is rejected at the given level of significance. In this example, we reject the null hypothesis and conclude that the observed investigation cycle time of 10.36 days after the training is significantly different than the original 20 days. Thus, it can be concluded that training has paid off because the investigation cycle time has been reduced significantly based on the training.

8.2.2 Two-Sample *t*-Test

If instead of comparing one group's mean against a target value we want to compare the means of two groups, then we can not use a one-sample *t*-test, but must use a two-sample *t*-test. As mentioned earlier, one assumption in performing a two-sample *t*-test is that the data follow a normal distribution. Prior to performing a two-sample *t*-test, we also need to evaluate the variances of each of the two groups. This is an important prerequisite because the data crunching process will be different for the case when the variances are assumed to be equal or when the variances are assumed to be unequal. To address the variances, we need to use an F-test or Bartlett's test, as will be discussed later. The following example will illustrate the concept of a two-sample t-test.

> A company wants to compare how long it takes (in hours) for each shift in the quality control laboratory to perform an analytical test on a specific product. There are two shifts: shift 1 and shift 2. A given test is selected, and the times to complete that test are collected and analyzed with statistical software.
>
> As mentioned, we need to evaluate each group's variance prior to performing the two-sample t-test. Figure 8.3 shows the results from Bartlett's test for equal variances. In this case, the null hypothesis is that all groups' variances are equal (or no significant difference exists between the variances), while the alternate hypothesis is that at least one group's variance is different. The same approach as for the comparison of the p-value with the significance level will be used. Remember that we will be using a significance level $\alpha = 0.05$. Figure 8.3 shows that shift 1 has a standard deviation of 0.6967, while shift 2 has a standard deviation of 0.4719. Are they significantly different? Taking a look at the p-value, we can observe that it is 0.0218. Using a significance

Bartlett's test for equal variance: duration (hours)
(Use with normal data)

Test information
H_0: Variance 1 = Variance 2 = . . . = Variance k
H_a: At least one pair Variance $i \neq$ Variance j

Shift	**1**	**2**
Count	31	42
Mean	3.526	4.306
Median	3.570	4.375
StDev	0.696743	0.471918
AD normality test p-value	0.4241	0.2521
Bartlett's test statistic	5.264	
p-value	**0.0218**	

Figure 8.3 Bartlett's test.

level of 0.05, we reject the null hypothesis and conclude that the variances are different.

Once we determine through Bartlett's test that the variances are not equal, we proceed to analyze the means of both groups. Remember that in the statistical software that you use for the analysis, you will have to select a check box specifiying "Assume unequal variances." Performing a two-sample *t*-test in a statistical software application, we obtain the results presented in Figure 8.4. Notice that the null hypothesis is that the mean difference is equal to zero. In other words, the null hypothesis is that the difference in the means is not significant. On the other hand, the alternate hypothesis is that the mean difference is not equal to zero; that is, the means are significantly different.

Figure 8.4 shows that the mean duration for the particular test for shift 1 is 3.526 hours, while the mean duration for shift 2 is 4.306 hours. Is that difference significant enough? Taking a look at the *p*-value, the analysis shows that it is again 0.0000 (not necessarily zero, but a very small value). Since the *p*-value is less than the required significance level, we say the null hypothesis is rejected at the given level of significance. In this example, we reject the null hypothesis and conclude that the observed difference in test duration for shift 1 versus shift 2 is significant. Thus, it can be concluded that shift 1 is completing the test in a significantly shorter time than shift 2.

Two-sample *t*-test: duration (hours)

Test information
H_0: Mean difference = 0
H_a: Mean difference ≠ 0
Assume unequal variance

Shift	**1**	**2**
Count	31	42
Mean	3.526	4.306
StDev	0.6967	0.4719
p-value (2-sided)	**0.0000**	

Figure 8.4 Two-sample *t*-test.

8.2.3 One-Way ANOVA Test

If, instead of comparing one group's mean against a target value (or comparing the means of two groups), we want to compare the means of three or more groups, we can use an *analysis of variance* (ANOVA) test. ANOVA is a statistical test that uses variances to compare multiple averages simultaneously. Instead of comparing pairwise averages, it compares the variance *between groups* with the variance *within groups*. The *between-group* variance is obtained from the variance of the group averages, while the *within-group* variance is obtained from the variance between values within each group and then pooled across the groups.

As mentioned earlier, one assumption in performing an ANOVA test is that the data follow a normal distribution. Prior to performing an ANOVA test, we also need to evaluate the variances of each of the groups. This is an important prerequisite because for the ANOVA test, the variances are assumed to be equal. To address the variances, we need to use an *F*-test or Bartlett test, as will be discussed later. The following example will illustrate the concept of an ANOVA test:

> A company wants to compare the performance of three shifts in terms of the manufacturing cycle time to determine if there are significant differences in their averages. After verifying the assumptions related to normality of the data and equal variances, the box plots shown in Figure 8.5 are developed. The dotted lines

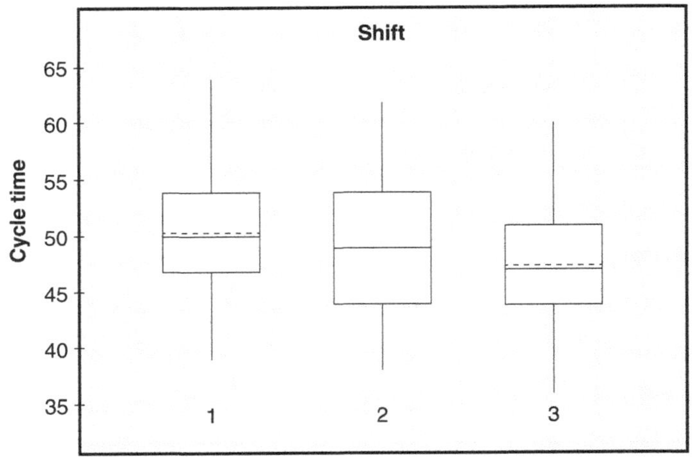

Figure 8.5 Box plots for manufacturing cycle time comparison example.

in the box plots represent the averages, while the solid lines represent the medians. It can be seen that, *within* each shift, the median and average are approximately the same. In a normal distribution, the average, median, and mode are similar. So, the data appear to follow the normal distribution. Also, the variability *between* the three shifts does not differ significantly. So, we can see graphically how the equal variances assumption is met. Furthermore, the averages *among* shifts also appear to be similar; that is, no significant difference can be observed between the three shifts. Although the averages apparently are not significantly different, an ANOVA test will be performed in order to obtain a statistical solution.

The ANOVA table shown in Figure 8.6 presents the averages for each of the three shifts: 50.26 minutes for shift 1, 49.12 minutes for shift 2, and 47.44 minutes for shift 3. Is there any difference between those averages? The null hypothesis for an ANOVA test is that all the averages are the same (no significant difference between the averages), while the alternate hypothesis is that at least one average is statistically different. As can be seen in Figure 8.6, the p-value for the ANOVA test is 0.2015. So, we conclude that there is not enough evidence to reject the null hypothesis. In other words, there is no statistical difference between the manufacturing cycle times for the three shifts.

One-way ANOVA and means matrix: cycle time

Test information

H_0: Mean 1 = Mean 2 = . . . = Mean k

H_a: At least one pair Mean $i \neq$ Mean j

Shift	**1**	**2**	**3**
Count	31	42	27
Mean	50.26	49.12	47.44
StDev	5.721	5.865	6.296

ANOVA table

Source	**SS**	**DF**	**MS**	**F**	***p*-value**
Between	114.95	2	57.477	1.629	0.2015
Within	3423.0	97	35.289		
Total	3538.0	99			

Figure 8.6 ANOVA test for manufacturing cycle time comparison example.

Let me illustrate the concept with another example:

> Although it was proven that the average manufacturing cycle time for each shift was not statistically different, the performance among the shifts does not appear to be the same. For that reason, the company developed a *quality index* to measure their performance. The index is based on a scale of 1 to 5, where 1 means poor performance and 5 means excellent performance. Figure 8.7 shows the box plots for the quality index of each shift. As with the manufacturing cycle times example, the averages and medians within each shift are very similar. So, the normality assumption can be observed graphically. However, as mentioned earlier, do not rely only on the graphical evaluation of the data; you need to perform an analytical evaluation of the data using the Anderson-Darling normality test or similar analysis. Also, variability between the shifts also appears to be very similar when analyzed graphically. But remember that in real life you will be analyzing the equality of variances through an F-test or Barlett test, not just through a graphical analysis. However, when we analyze graphically the averages among the three shifts, certain differences can be observed, as shown in Figure 8.7.
>
> Taking a look at the box plots in Figure 8.7, a higher value for the average of shift 2 is noticeable. On the other hand, the averages for shift 1 and shift 3 do not differ so much. The statistical

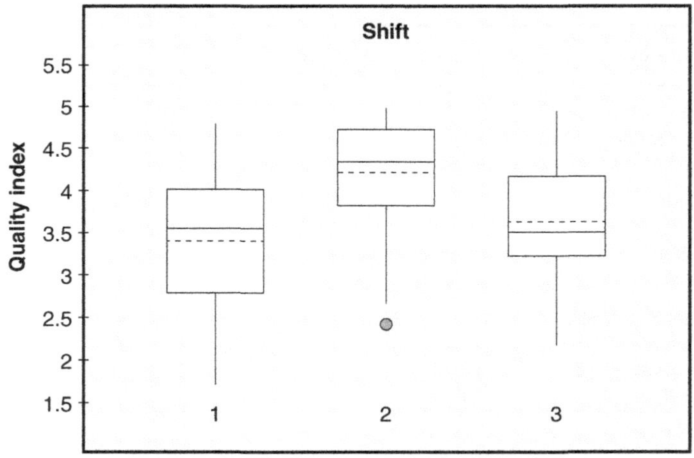

Figure 8.7 Box plots for quality index comparison example.

One-way ANOVA and means matrix: quality index

Test information
H_0: Mean 1 = Mean 2 = . . . = Mean *k*
H_a: At least one pair Mean *i* ≠ Mean *j*

Shift	**1**	**2**	**3**		
Count	31	42	27		
Mean	3.39	4.21	3.64		
StDev	0.825	0.621	0.670		
ANOVA table					
Source	**SS**	**DF**	**MS**	***F***	***p*-value**
Between	12.700	2	6.350	12.856	**0.0000**
Within	47.912	97	0.493943		
Total	60.612	99			

Figure 8.8 ANOVA test for quality index comparison example.

significance of this difference will be analyzed through an ANOVA test. The table shown in Figure 8.8 presents the quality index averages for each of the three shifts: 3.39 for shift 1, 4.21 for shift 2, and 3.64 for shift 3. Is there any difference between these averages? The null hypothesis for an ANOVA test is that all the averages are the same (no significant difference between the averages), while the alternate hypothesis is that at least one average is statistically different. As can be seen in Figure 8.8, the *p*-value for the ANOVA test is 0.0000. So, we reject the null hypothesis. In other words, there is statistical difference between the quality indexes for the three shifts.

8.2.4 Two-Way ANOVA Test

What if, instead of comparing different groups of one factor (for example, machine), we want to compare different groups of two factors (for example, machine *and* material). In this case, we can use the two-way ANOVA. This analysis is an extension to the one-way ANOVA. There are two independent variables (hence the name two-way ANOVA). In order to perform this type of analysis, certain assumptions must be met: (1) the populations from which the samples were obtained must be normally or approximately normally distributed; (2) the samples must be independent; (3) the variances of the populations must be equal; and (4) the groups must have the same sample size.

There are three sets of hypotheses with the two-way ANOVA. The null hypotheses for each of the sets are:

- The population means of the first factor are equal. This is like the one-way ANOVA for the *row* factor.
- The population means of the second factor are equal. This is like the one-way ANOVA for the *column* factor.
- There is no interaction between the two factors. This is similar to performing a test for independence.

Let us explain the two-way ANOVA with an example:

A company is evaluating certain suppliers of a specific material to be used in their molding process. There are three suppliers for that material. Moreover, the company has four machines in which the parts can be molded, each machine from a different manufacturer. So, the company would like to learn if there are significant differences in the suppliers (materials) and the machines. In order to analyze this, each material is used in each machine to produce a molded part. Then, a two-way ANOVA is performed. The first factor is *machine*, while the second factor is *material*. Figure 8.9 shows the box plots for machines and materials.

The specification for the weight of the molded part ranges from 1.5 to 2.5 g. The box plots for machine in Figure 8.9 show that the weights for machine 1, machine 2, and machine 3 behave in a similar way; that is, their average and variation are very similar. However, although the variation for machine 4 seems similar to that of the other machines, its average goes above the upper specification limit. When analyzing the box plots for material (supplier)

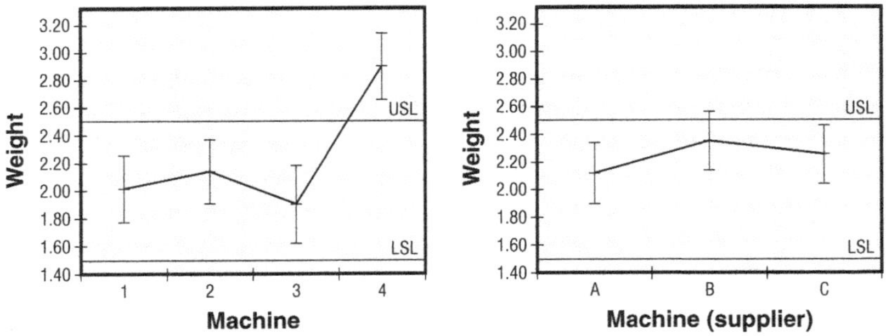

Figure 8.9 Box plots for machine and material example.

in Figure 8.9, the weights' averages and variability look very similar. Figure 8.10 shows the two-way ANOVA for this example.

The null hypothesis for the first factor (machine) is that their averages are not significantly different, whereas the alternate hypothesis is that their averages are significantly different. Since the *p*-value for *machine* is lower than the alpha value of 0.05 (*p*-value = 0.0000), we reject the null hypothesis and conclude that at least one of the machine's averages is different. Guess which one? Machine 4, of course. What about the material's averages? The null hypothesis for the second factor (material) is that their averages are not significantly different, whereas the alternate hypothesis is that their averages are significantly different. Since the *p*-value for *material* is higher than the alpha value of 0.05 (*p*-value = 0.4296), we do not have evidence to reject the null hypothesis. Data do not show there is statistical difference between the averages for the materials. However, from a practical point of view, the box plots for material in Figure 8.9 show that material A is the one that produced a part that is more centered within the specification limits. So, from a business perspective, material A must be selected.

However, there is another aspect we need to consider: the interactions. As will be seen in Chapter 10, an interaction occurs

Two-way ANOVA: weight

Test information

H_0 (factor machine): Mean 1 = Mean 2 = . . . = Mean *k*
H_a (factor machine): At least one pair Mean $i \neq$ Mean j

H_0 (factor material [supplier]): Mean 1 = Mean 2 = . . . = Mean *k*
H_a (factor material [supplier]): At least one pair Mean $i \neq$ Mean j

H_0 (interaction): There is no interaction between factors X1 and X2
H_a (interaction): There is an interaction between factors X1 and X2

ANOVA table

Source	**DF**	**SS**	**MS**	***F***	***p*-value**
Machine	3	11.951	3.984	12.960	**0.0000**
Material (supplier)	2	0.526059	0.263030	0.855689	0.4296
Interaction	6	1.295	0.215909	0.702395	0.6486
Error	67	20.595	0.307389		
Total	78	34.821	0.446418		

Figure 8.10 Two-way ANOVA test for machine and supplier example.

when the behavior of one factor may be dependent on the level of another factor. In our example, an interaction would exist if the result depends on which combination of machine and material is used. The null hypothesis for the interaction is that it is not significant, whereas the alternate hypothesis is that it is significant. Since the p-value for the interaction is higher than the alpha value of 0.05 (p-value = 0.6486), we do not have evidence to reject the null hypothesis. Data do not show that the interaction of machines and materials is significant. In summary, the company should purchase material A and could produce the part in machine 1, machine 2, or machine 3.

8.3 COMPARING MEDIANS

So far, we have been performing hypothesis tests for data that follow the normal distribution. But what if the data do not follow a normal distribution? In that case, we can either transform the data or run a *nonparametric* test, as discussed in Section 5.9. We will explain the latter approach in this chapter. The three nonparametric tests that will be explained are the one-sample sign test, two-sample Mann-Whitney test, and Kruskal-Wallis test.

8.3.1 One-Sample Sign Test

The *one-sample sign test* is a nonparametric equivalent to the one-sample t-test. The nonparametric tests are performed whenever the data do not follow the normal distribution. Recall that whenever the data follow the normal distribution, we can use either the average, the median, or the mode as the measure of central tendency because all of them are very similar. However, when the data do not follow the normal distribution (that is, the data are skewed), then we need to use the median instead of the average because the median is less impacted by outliers than the average.

Let us illustrate the use of the one-sample sign test with an example:

> The median cycle time for investigation has been 25 days (50% of investigations needed more than 25 days to be completed). After new root cause analysis training, three months of data were used to evaluate the improvement (if any). Figure 8.11 shows the results of the one-sample sign test using statistical software.
>
> It can be noted that the median for the investigation cycle time of the 100 samples taken after the training's completion was reduced from 25 days to 9 days. The question might be "Is the

One-sample sign test	
Test information	
H_0: Median = 25	
H_a: Median less than 25	
Results	**Days after**
Count (N)	100
Median	9
Points below 25	93
Points equal to 25	1
Points above 25	6
p-value (1-sided)	**0.0000**

Figure 8.11 One-sample sign test example.

observed difference the result of the training or just the result of common cause variation?" The null hypothesis was established as a median equal to 25 days, while the alternate hypothesis was established as a median of less than 25 days, which is what we would expect. Recall from Section 8.1 that the criterion to determine if the null hypothesis will be rejected or not is going to be the *p*-value. In that section I mentioned that "*if* p-*value is low,* H_0 *must go.*" In our examples, for consistency purposes, I will use a producer's confidence of 95%, which means that $\alpha = 0.05$, or a significance level of 5%.

Taking a look at Figure 8.11, the obtained *p*-value = 0.0000 (not necessarily zero, but a very small number). We already mentioned that if the *p*-value is less than the required significance level, then we say the null hypothesis is rejected at the given level of significance. In this example, we reject the null hypothesis and conclude that the observed investigation cycle time of 9 days after the training is significantly different than the original 25 days. Thus, it can be concluded that training has paid off because the investigation cycle time has been reduced significantly based on the training.

8.3.2 Two-Sample Mann-Whitney Test

The *two-sample Mann-Whitney test* is a nonparametric equivalent to the two-sample *t*-test. The nonparametric tests are performed whenever the data do not follow the normal distribution. The two-sample Mann-

Whitney test compares the medians of two groups. Let us illustrate the use of this test with an example:

> A pharmaceutical company wanted to determine if there are significant differences in the hardness of the tablet they manufacture using two different pieces of equipment. In order to analyze this, a period of one year of production was collected from each machine. After performing a normality test, it was noticed that data do not follow the normal distribution. The histograms in Figure 8.12 show a *p*-value for the Anderson-Darling normality test of 0.0000 for each machine, indicating that the data do not follow the normal distribution.
>
> Since the data do not follow a normal distribution, a two-sample Mann-Whitney test was performed with the data to compare the two medians. Figure 8.13 shows the results of the test, in which a *p*-value of 0.0000 was obtained. A *p*-value lower than 0.05 for this test indicates that the medians of each of these popu-

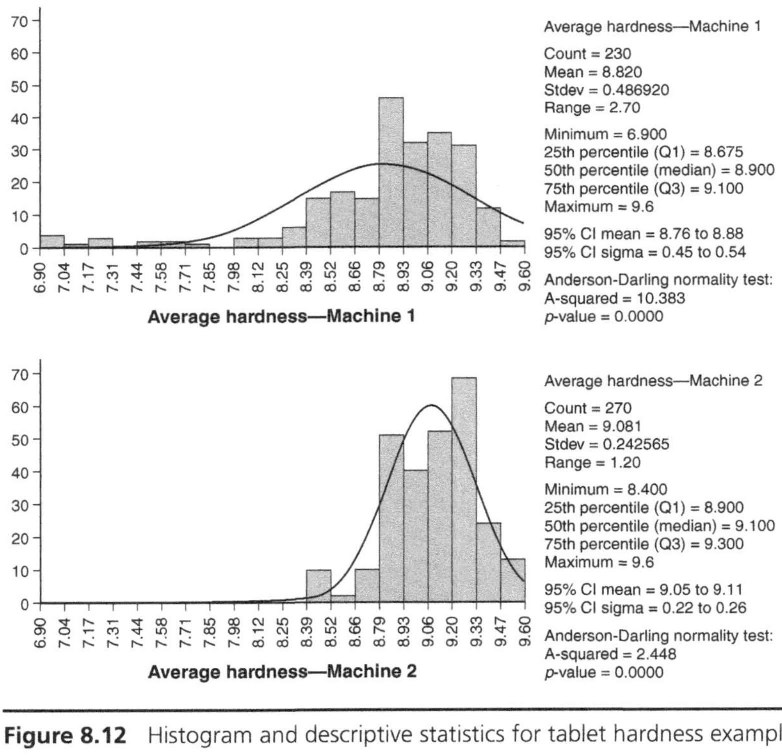

Figure 8.12 Histogram and descriptive statistics for tablet hardness example.

Two-sample Mann-Whitney test: average hardness		
Test information		
H_0: Median difference = 0		
H_a: Median difference ≠ 0		
Machine	**1**	**2**
Count	230	270
Median	8.90	9.10
Mann-Whitney statistic	46569.50	
p-value (2-sided, adjusted for ties)	**0.0000**	

Figure 8.13 Two-sample Mann-Whitney test for table hardness example.

lations are statistically different. The obtained value indicates that the difference in the median (8.90 kp for machine 1 and 9.10 kp for machine 2) is due to some special cause(s). Such difference in the medians is not the result of natural process variability. So, it must be investigated why machine 2 is providing a higher hardness than machine 1.

8.3.3 Kruskal-Wallis Test

The *Kruskal-Wallis test* is a nonparametric equivalent to the one-way ANOVA test. The nonparametric tests are performed whenever the data do not follow the normal distribution. The Kruskal-Wallis test is used for comparing more than two samples that are independent, or not related. The null hypothesis is that the populations from which the samples originate have the same median. When the Kruskal-Wallis test leads to significant results, then at least one of the samples is different from the other samples. The test does not identify where the differences occur or how many differences actually occur. It is an extension of the two-sample Mann-Whitney test for comparing three or more groups. The two-sample Mann-Whitney test would help analyze the specific sample pairs for significant differences.

Since it is a nonparametric method, the Kruskal-Wallis test does not assume a normal distribution, unlike the analogous one-way ANOVA. However, the test does assume an identically shaped and scaled distribution for each group, except for any difference in medians.

Let us explain the Kruskal-Wallis test through an example:

A pharmaceutical company wanted to analyze three different suppliers of a fluid used in their manufacturing process in order to select one of them. A critical characteristic of the fluid is viscosity, measured in *centipoise* (cP). The specifications for the viscosity of the fluid range from 2.4 to 3.6 cP. Data from the three suppliers were analyzed and a histogram and descriptive statistics of them generated. Figure 8.14 shows the results for the three

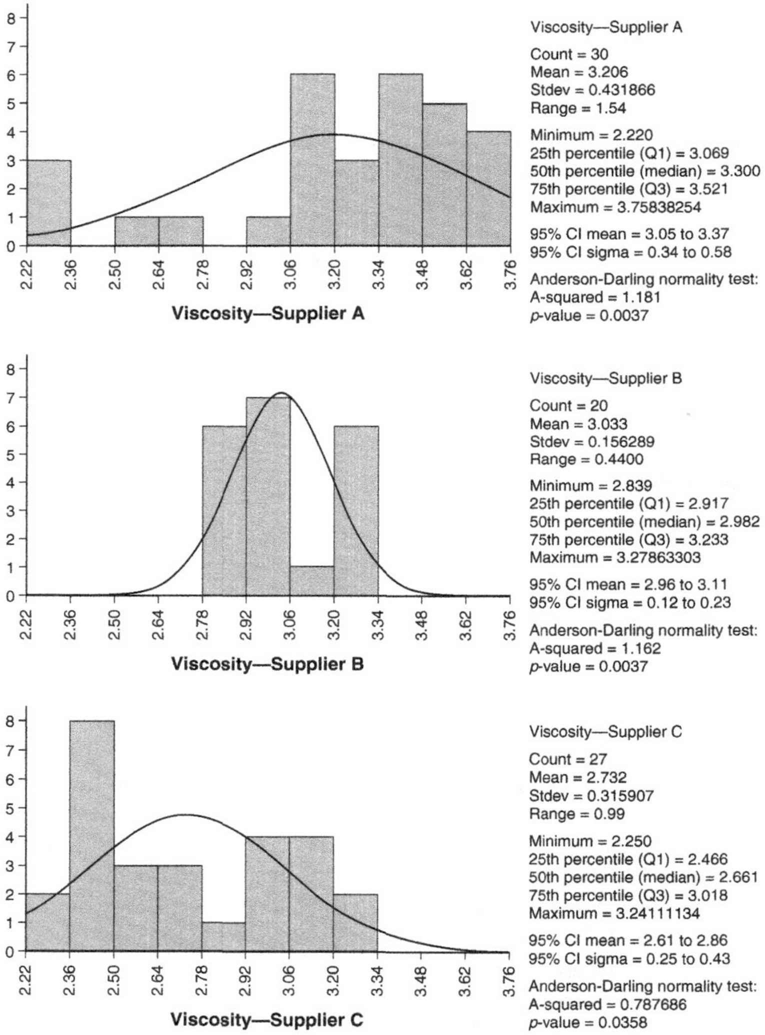

Figure 8.14 Histogram and descriptive statistics for viscosity example.

suppliers. From the Anderson-Darling normality test, it can be seen that none of the three data sets follow a normal distribution because the *p*-values for them are below 0.05, which was the alpha level (α) selected by the company. Since the data do not follow the normal distribution, we will use the median instead of the mean as the measure of central tendency, as mentioned in Section 4.5. It can be seen that the median for supplier A is 3.30 cP, for supplier B, 2.98 cP, and for supplier C, 2.66 cP.

In order to analyze the medians, a Kruskal-Wallis test will be performed. The null hypothesis is that the medians are not significantly different, while the alternate hypothesis is that at least one median is different. Figure 8.15 shows the results of the Kruskal-Wallis test. A *p*-value of 0.0000 was obtained. A *p*-value lower than 0.05 for this test indicates that the medians of each of these populations are statistically different. The obtained value indicates that the difference in the medians (3.30 cP for supplier A, 2.98 cP for supplier B, and 2.66 cP for supplier C) is due to some special cause(s). Such differences in the medians are not the result of natural process variability.

Since supplier B has a median of 2.98 cP, which is closest to the target value of 3.00 cP, that supplier must be selected. It can also be noted that the data from supplier B have the smallest standard deviation among the three suppliers, as can be seen in Figure 8.14. However, the equality of variances when the data are not normal will be discussed in Section 8.4.3 about the Levene test.

Kruskal-Wallis nonparametric ANOVA: viscosity

Test information
H_0: Median 1 = Median 2 = . . . = Median *k*
H_a: At least one pair Median $i \neq$ Median j

Supplier	**A**	**B**	**C**
Count (N)	30	20	27
Median	3.30	2.98	2.66
p-value (2-sided, adjusted for ties)	**0.0000**		

Figure 8.15 Kruskal-Wallis test for viscosity example.

8.4 COMPARING VARIANCES

Sections 8.2 and 8.3 were about the comparison of means and medians, respectively. Those are measures of central tendency, as discussed in Section 4.4. However, as discussed in Section 7.3 about process capability, sometimes our issues are not about process *centering* but about process *dispersion*. One of the measures of dispersion discussed in Section 4.4 was the *variance*. In this section we will discuss three hypothesis tests for comparing variances: the *F*-test, the Bartlett test, and the Levene test. The first two tests will be used when the data follow the normal distribution, whereas the last test will be used for nonnormal data.

8.4.1 *F*-Test

The *F*-test is used whenever we want to compare the variances between two groups. The underlying assumption is that the data follow a normal distribution. As mentioned in Section 8.2.3, this test has to be performed prior to a one-way ANOVA because, in order to perform such a test, the variances must be equal. Let us illustrate the use of the *F*-test with an example:

> A company wants to analyze two different adhesives for the transdermal patches they manufacture. Adhesives on a patch generally help maintain contact between the transdermal system and skin surface. The adhesiveness of the patches is critical in the drug delivery mechanism, its safety, product quality, and efficacy. As such, a good adhesive should easily adhere to the skin with an applied finger pressure and be tacky enough to maintain a strong holding force. The adhesive should also be easily removed from the skin without leaving a residue. The specification for the adhesiveness of the transdermal patch ranges from 300 to 3500 g/system. The box plots for the transdermal patches using each adhesive are presented in Figure 8.16.
>
> It can be seen that, although the averages seem to be similar, the variation in adhesiveness of adhesive A is smaller than that of adhesive B. In order to analyze the variances of these two groups, an *F*-test will be peformed using statistical software. The results of the *F*-test are summarized in Figure 8.17.
>
> As can be seen in Figure 8.17, the null hypothesis is that the variances are not significantly different, while the alternate hypothesis is that they are different. Since the *p*-value is lower than 0.05, we reject the null hypothesis and conclude there are significant differences in the variances. Adhesive A has a standard

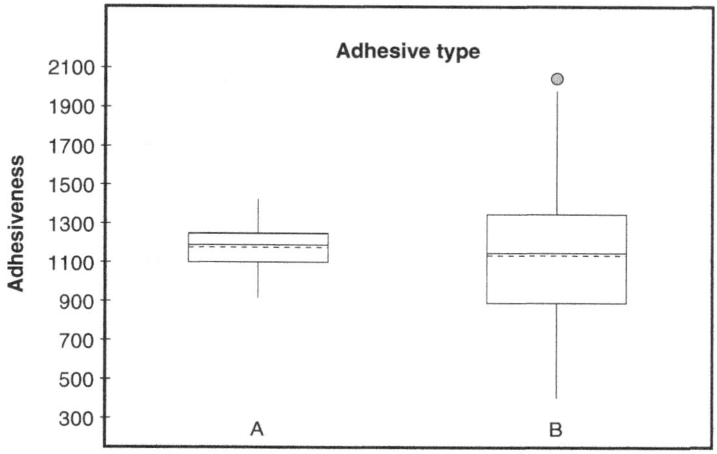

Figure 8.16 Box plots for transdermal patch adhesiveness example.

F-test for equal variance: adhesiveness
(Use with normal data)

Test information
H_0: Variance 1 = Variance 2
H_a: Variance 1 ≠ Variance 2

Adhesive type	**A**	**B**
Count	100	125
Mean	1179.2	1136.0
StDev	102.62	335.57
AD normality test p-value	0.1975	0.7584
p-value	**0.0000**	

Figure 8.17 F-test for transdermal patch adhesiveness example.

deviation of 102.6 g/system, while adhesive B has a standard deviation of 335.6 g/system. Since the averages for both adhesives are about the same (1179.2 for adhesive A and 1136.0 for adhesive B), the adhesive with the lower variation must be selected. In this case, adhesive A must be the chosen one.

At this point, we might ask the question of whether the F-test was the appropriate one for these data since we did not perform a normality test prior to performing this test. However, as can be seen in Figure 8.17, an Anderson-Darling normality test was performed as part of the analysis. The p-values for the normality test are higher than 0.05; thus, we can not reject the null hypothesis that the data follow a normal distribution. If any of those p-values were lower than 0.05, we could not have chosen the F-test (or Bartlett test, for three or more variances). In such case, a Levene test would be the appropriate one.

8.4.2 Bartlett Test

The Bartlett test is used whenever we want to compare the variances between two or more groups. The underlying assumption is that the data follow a normal distribution. As mentioned in Section 8.2.3, this test has to be performed prior to a one-way ANOVA because, in order to perform such a test, the variances must be equal. Let us illustrate the use of the Bartlett test with an example:

Let us suppose that in the previous example the manufacturer would like to test a third adhesive. Since the F-test can only compare two variances at a time, we need to perform a Bartlett test. Figure 8.18 shows the results for the adhesiveness tests.

It can be seen that, although the averages seem to be similar, the variation in adhesiveness of adhesive A is smaller than that

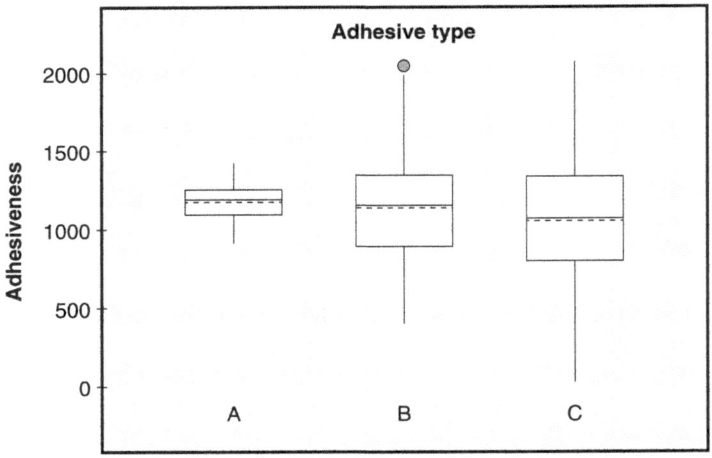

Figure 8.18 Box plots for transdermal patch adhesiveness example.

of adhesive B and adhesive C. In order to analyze the variances of these three groups, a Bartlett test will be peformed using statistical software. The results of the Bartlett test are summarized in Figure 8.19.

As can be seen in Figure 8.19, the null hypothesis is that the variances are not significantly different, while the alternate hypothesis is that at least one of the variances is different. Since the p-value is lower than 0.05, we reject the null hypothesis and conclude there are significant differences in the variances. Adhesive A has a standard deviation of 102.6 g/system, adhesive B has a standard deviation of 335.6 g/system, and adhesive C has a variance of 380.6 g/system. Since the averages for the three adhesives are about the same (1179.2 for adhesive A, 1136.0 for adhesive B, and 1057.2 for adhesive C), the adhesive with the lower variation must be selected. As in the example where the F-test was used, adhesive A must be the chosen one.

Another graph will be developed for this analysis: the histogram and descriptive statistics. The results for each of the three adhesives are presented in Figure 8.20.

The Anderson-Darling normality test shows that the normality hypothesis can not be rejected for any adhesive type. Furthermore, a look at the minimum values shows that adhesive C will not be complying with the lower specification limit of 300 g/system because there are values below that specification limit.

Bartlett test for equal variance: adhesiveness
(Use with normal data)

Test information
H_0: Variance 1 = Variance 2 = . . . = Variance k
H_a: At least one pair Variance $i \neq$ Variance j

Adhesive type	**A**	**B**	**C**
Count	100	125	200
Mean	1179.2	1136.0	1057.2
StDev	102.62	335.57	380.64
AD normality test p-value	0.1975	0.7584	0.2818
p-value	**0.0000**		

Figure 8.19 Bartlett test for transdermal patch adhesiveness example.

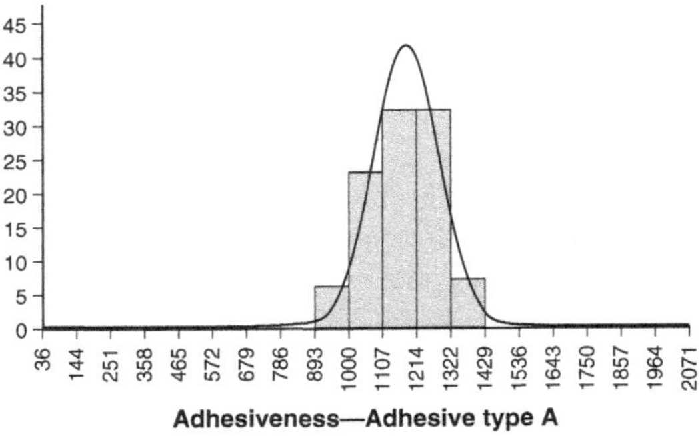

Adhesiveness—Adhesive type A

Count = 100
Mean = 1179.2
Stdev = 102.62
Range = 494.79

Minimum = 923.56
25th percentile (Q1) = 1100.0
50th percentile (median) = 1191.0
75th percentile (Q3) = 1253.3
Maximum = 1418.35089

95% CI mean = 1158.82 to 1199.54
95% CI sigma = 90.10 to 119.21

Anderson-Darling normality test:
A-squared = 0.505843
p-value = 0.1975

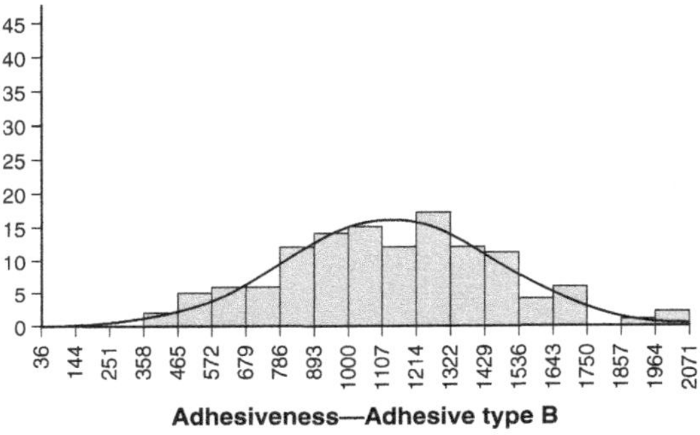

Adhesiveness—Adhesive type B

Count = 125
Mean = 1136.0
Stdev = 335.57
Range = 1648.18

Minimum = 404.40
25th percentile (Q1) = 894.03
50th percentile (median) = 1155.6
75th percentile (Q3) = 1351.0
Maximum = 2052.57371

95% CI mean = 1076.56 to 1195.38
95% CI sigma = 298.50 to 383.24

Anderson-Darling normality test:
A-squared = 0.244274
p-value = 0.7584

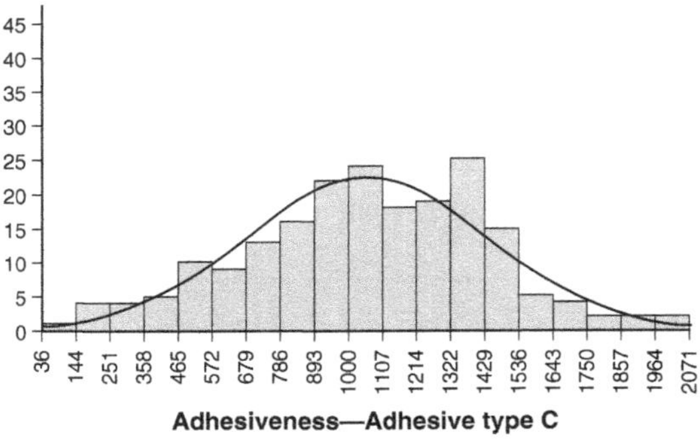

Adhesiveness—Adhesive type C

Count = 200
Mean = 1057.2
Stdev = 380.64
Range = 2032.76

Minimum = 36.458
25th percentile (Q1) = 804.88
50th percentile (median) = 1070.1
75th percentile (Q3) = 1330.6
Maximum = 2069.22192

95% CI mean = 1004.14 to 1110.29
95% CI sigma = 346.63 to 422.10

Anderson-Darling normality test:
A-squared = 0.444532
p-value = 0.2818

Figure 8.20 Histogram and descriptive statistics for transdermal patch adhesiveness example.

8.4.3 Levene Test

The F-test and Bartlett test are used when the data follow a normal distribution. When the data do not follow a normal distribution, we can perform the Levene test. As opposed to the F-test and Bartlett test, the Levene test applies for the comparison of two or more variances. Let us illustrate the use of the Levene test through an example:

> A pharmaceutical company wants to evaluate the time it takes their analysts to perform a certain laboratory test. They have noticed too much variation in the time it takes the various shifts to complete the test and would like to investigate the reason for the variation. In order to start the analysis, the time it took each analyst to perform the laboratory test (in hours) is collected. Then, the data are segregated by shift. The box plots for the three shifts are presented in Figure 8.21.
>
> It can be seen that although shift 2 takes a longer time to complete the test, the analysts in that shift are more consistent; in other words, their variability is much less than the variability of shift 1 and shift 3. Furthermore, there are three outliers that obtained shorter times than the rest of shift 2. Figure 8.22 shows the Anderson-Darling normality test results in terms of their p-values: 0.0021 for shift 1, 0.0000 for shift 2, and 0.0190 for shift 3. Based

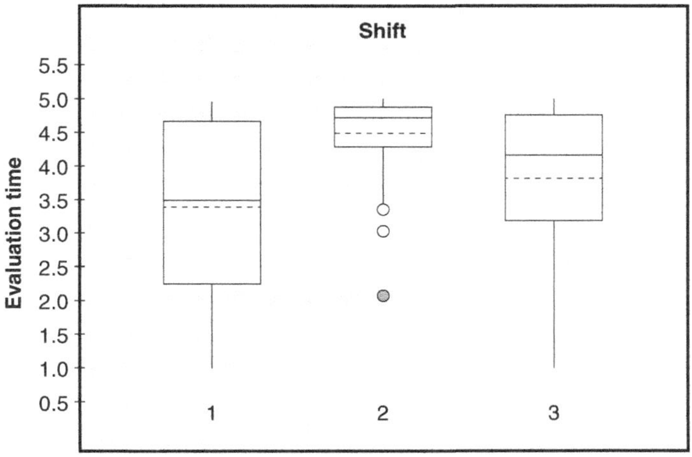

Figure 8.21 Box plots for laboratory test evaluation example.

Levene test for equal variance: evaluation time
(Use with nonnormal data)

Test information
H_0: Variance 1 = Variance 2 = . . . = Variance *k*
H_a: At least one pair Variance $i \neq$ Variance j

Shift	**1**	**2**	**3**
Count	31	42	27
Mean	3.41	4.52	3.82
Median	3.50	4.75	4.18
StDev	1.304	0.618	1.092
AD normality test *p*-value	**0.0021**	**0.0000**	**0.0190**
p-value	**0.0000**		

Figure 8.22 Levene test for laboratory test evaluation example.

on these results, we can conclude that none of the data sets follow the normal distribution. So, as mentioned earlier, we can not use the *F*-test or Bartlett test to analyze the variances. Instead, we need to use the Levene test to compare the variances. Figure 8.22 shows the variances for the three shifts: 1.304 hours for shift 1, 0.618 hours for shift 2, and 1.092 hours for shift 3.

In the Levene test, the null hypothesis is that there is not a significant difference in the variances, and the alternate hypothesis is that at least one variance is significantly different. Taking a look at the *p*-value for that test (0.0000), we can conclude that at least one variance is significantly different. Analyzing the results, it is evident that the variation in shift 2 is much lower than the variation in shift 1 and shift 3. So, the company will need to identify the causes of the smaller variation in that shift as compared with the other two shifts.

8.5 SUMMARY

Whenever we want to make improvement in our processes, some of the most useful tests for analyzing different alternatives are the hypothesis tests. Using this tool, several groups can be compared in terms of averages, medians, variances, and so on. In a hypothesis test, we want to determine whether the observed difference between groups is due to the random variation of the process or if that difference is caused by a change in the process.

Prior to performing a hypothesis test, we need to address the normality of the data: if data follow a normal distribution, the tests to be performed are called *parametric* tests; if data do not follow a normal distribution, the tests are called *nonparametric* tests. The parametric tests for central tendency use the *average*, while the nonparametric tests for central tendency use the *median*. When comparing central tendency, we need to select the test based on the number of groups we want to compare: a group against a standard, a group against another group, or more than two groups. The results of the hypothesis tests will be used in subsequent analyses.

9

Regression Analysis

9.1 OVERVIEW

Regression analysis is a statistical technique for estimating the relationships between variables. The focus is on the relationship between a *dependent* variable (also known as *output* or *y*) and one or more *independent* variables (also known as *input* or *x*). Specifically, regression analysis helps us to understand how the typical value of the dependent variable changes when any one of the independent variables is changed while the other independent variables are held fixed. A *scatter plot* shows the correlation between two variables in a process. Dots representing data points are scattered on the diagram. The extent to which the dots cluster together in a line across the diagram shows the strength with which the two factors are related. If the variables are correlated, when one changes, the other probably also changes. Dots that look like they are trying to form a line are strongly correlated. Figure 9.1 shows the different types of linear correlation than can be observed in a process.

Regression analysis is widely used for prediction and forecasting. It is also used to understand which of the independent variables are related to the dependent variable, and to explore the forms of these relationships. In some cases, regression analysis can be used to infer causal relationships between the independent and dependent variables. However, this can lead to false relationships, so caution is advisable. In other words, correlation does not imply causation. For example, although a scatter plot could show a positive correlation between the amount of rain and the amount of wealth, there is no causation between those two variables.

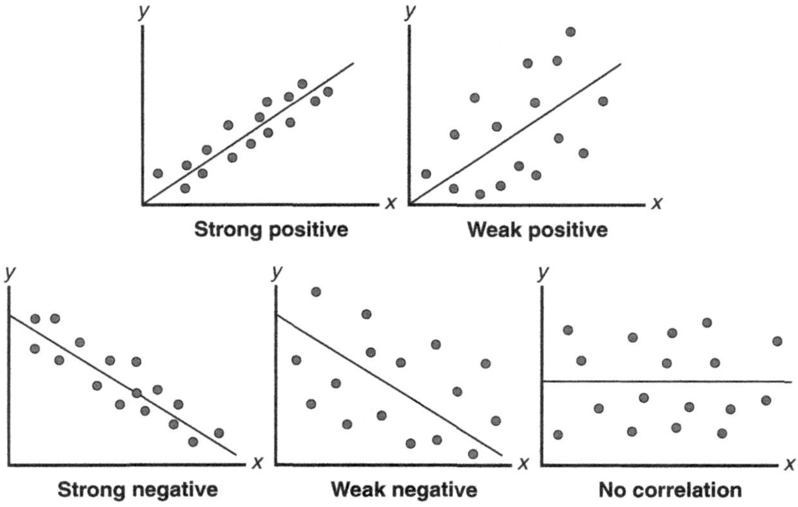

Figure 9.1 Types of correlation.

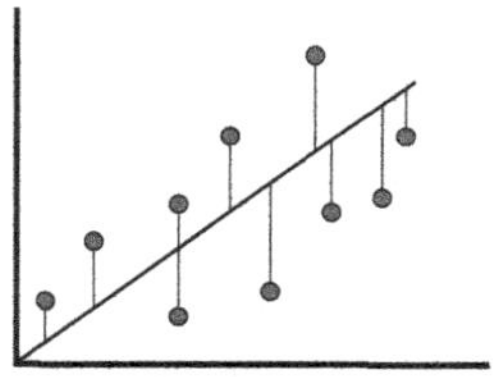

Figure 9.2 The least squares method.

9.2 LEAST SQUARES METHOD

The regression equation is determined by a procedure that minimizes the total squared distance of *all* points to the regression line. This procedure is called the *least squares method.* It finds the line where the squared vertical distance from each data point to the regression line (called *residuals*) is as small as possible. Regression uses the least squares method to determine the "best fitting line." In other words, the principle of least squares is "Choose, as the best fitting line, the line that minimizes the sum of squares of the deviations of the observed values of *Y* from those predicted." Figure 9.2 shows the concept behind the least squares method.

9.3 REGRESSION METRICS

Once the "best" regression line is obtained, we need to analyze some metrics to determine how appropriate the regression model is. Two of the metrics that will be discussed are the *correlation coefficient* and the *determination coefficient.*

The correlation coefficient (r) is a number that ranges from –1 to +1. This metric provides two important aspects of the regression model: *magnitude* and *slope direction.* For instance, when the correlation coefficient is equal to –1, all the data points fall in a straight, decreasing line. This is also called a "perfect negative correlation." When the correlation coefficient is equal to +1, all the data points fall in a straight, increasing line. This is also called a "perfect positive correlation." Figure 9.3 illustrates this concept.

When the data points begin to scatter away from the straight line, the correlation coefficient starts to move away from –1 or +1. That is, they move from those extreme values to zero. A correlation coefficient of zero means that there is no linear correlation. In summary, the closer the value to either –1 or +1, the stronger the correlation; the closer to zero, the weaker the correlation. Figure 9.4 shows this concept.

The other metric is called the *determination coefficient.* It is often referred to as r^2 or R^2. It is basically the square of the correlation coefficient. Based on the values mentioned regarding the correlation coefficient (range from –1 to +1), the range of values for the determination coefficient goes from zero to +1 (or from 0% to 100%, when expressed as a percentage). But what specific information provides the determination coefficient? This value represents the percentage of the model defined by the regression equation. That is, if the correlation coefficient is either –1 or +1, then the determination coefficient will be +1 (or 100%). That would represent a perfect correlation. In other words, the regression line would represent 100%

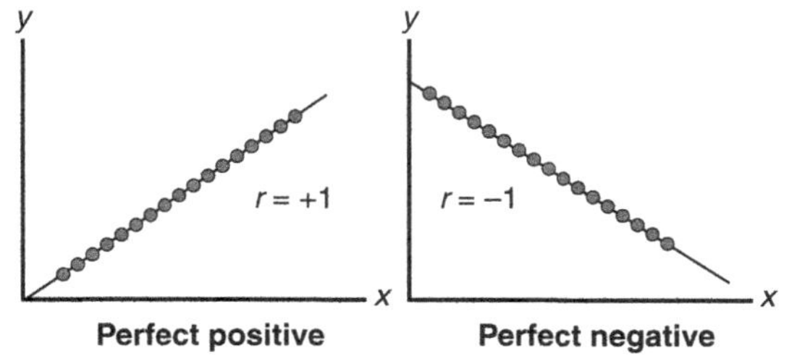

Figure 9.3 Perfect correlation.

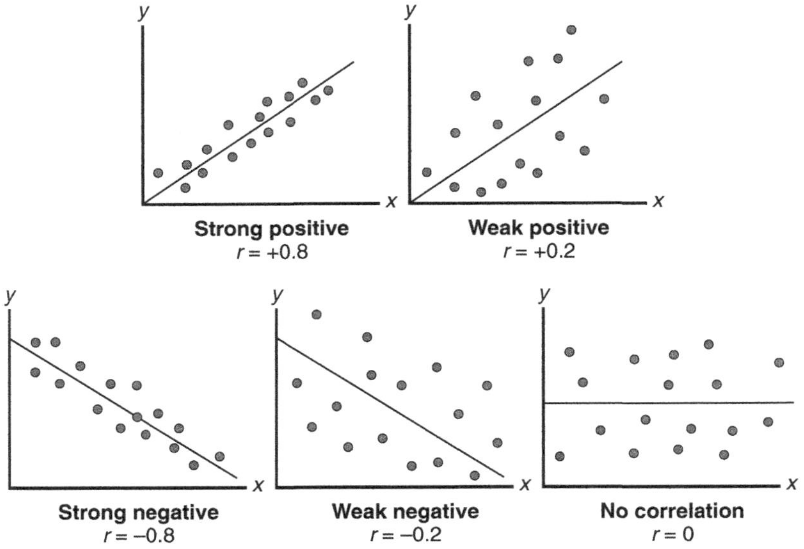

Figure 9.4 Strong and weak correlation.

of the model. Said differently, the independent variable would account for all the variation in the model.

However, when the correlation coefficient starts to move away from either –1 or +1, there are other independent variables that could be causing the variation. For instance, if the correlation coefficient is –0.8 or +0.8, the determination coefficient will be 0.64, or 64%, which means that the regression line only accounts for 64% of the model. In other words, there is 36% of the variation not explained by the regression equation. In such a case, an approach would be to identify other potential variables that could be affecting the model and include them in a multiple regression model. So, what is an "appropriate" value for the determination coefficient? There is not any specific agreement; however, most references establish that an R^2 greater than 0.80, or 80%, is considered acceptable.

9.4 RESIDUALS ANALYSIS

Once we have determined the best regression line equation and calculated the correlation and determination coefficient, there is another task we need to perform in order to make certain that the calculated regression line equation is statistically valid. This is one task that is rarely performed but is of paramount importance. It is called *residuals analysis*. As mentioned in Section 9.2, the residuals are the vertical distance from each data point to the regression line. Said differently, the residuals are the difference between the *expected* value and the *observed* value. The expected value is the value that is represented by the regression line, while the observed value is the actual value observed in the process. In order for the regression analysis to be statistically valid, the residuals must comply with the following assumptions:

- Residuals must be normally distributed.
- Residuals must be in statistical control.
- The average of the residuals must be zero.
- The variation of the residuals must remain constant.

There are many graphical tools that can be used to analyze these assumptions. Most statistical software has menus to analyze the residuals. However, those graphs can be prepared individually. For instance, to analyze the assumption of normality, we could use a histogram and descriptive statistics, as presented in Section 5.2. Also, in order to verify the assumptions of statistical control, average of zero, and constant variation, a variables control chart (as will be explained in Section 11.2) can be used. Figure 9.5 shows the histogram and descriptive statistics for the residuals analysis of a certain process. By looking at both the shape of the histogram and at the p-value for the Anderson-Darling normality test (0.8869), it can be seen that the normality assumption for residuals is achieved.

In order to verify the other three assumptions, an individuals control chart of the residuals is developed. It can be seen in Figure 9.6 that all data points (residuals) are within the control limits; that is, residuals are in

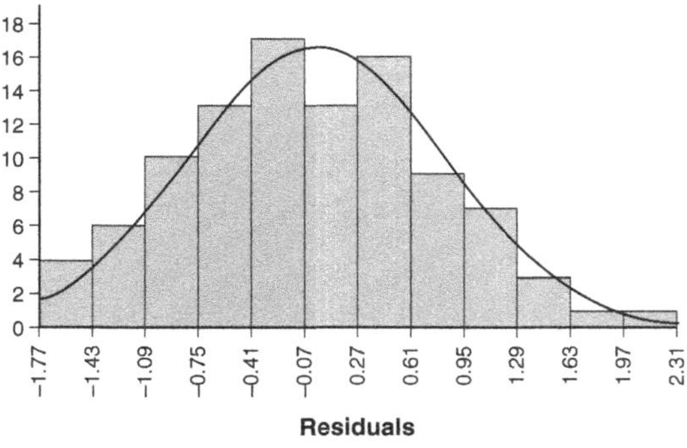

Residuals

Count = 100
Mean = −4.827E-15
Stdev = 0.820557
Range = 4.07

Minimum = 1.767
25th percentile (Q1) = −0.509577
50th percentile (median) = −0.0501819
75th percentile (Q3) = 0.570971
Maximum = 2.30278772

95% CI mean = −0.16 to 0.16
95% CI sigma = 0.72 to 0.95

Anderson-Darling normality test:
A-squared = 0.196609
p-value = 0.8869

Figure 9.5 Histogram and descriptive statistics for residuals analysis.

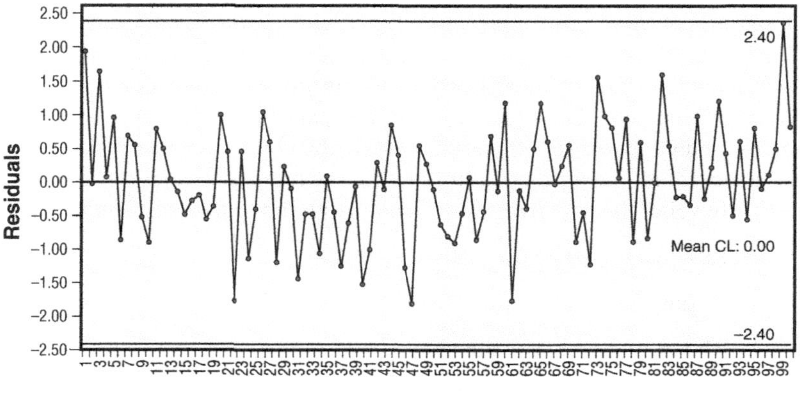

Figure 9.6 Individuals control chart for residuals analysis.

statistical control. It can also be noted that the average of the residuals (the centerline of the graph) is zero. The control chart also shows that variation has remained constant over time (that is, it does not show a pattern of increasing or decreasing over time).

Now that we know the basics of regression analysis, let us explore two types of regression analysis: simple linear regression and multiple linear regression.

9.5 SIMPLE LINEAR REGRESSION

In *simple linear regression*, we analyze the relationship between one independent variable and one dependent variable. The equation for the simple linear regression model is:

$$y = \beta_0 + \beta_1 x + \varepsilon$$

where y is the dependent variable, x is the independent variable, β_0 is the y-intercept, and β_1 is the slope. There is also an error term, defined by ε. Let us explain the concept with an example:

> A medical device company manufactures dental floss. Their customer satisfaction department wants to analyze which are the variables that have the most significant impact on customer satisfaction. So, after a customer complaint was received at their call center and finally resolved, certain information was requested from the complainant. The information collected included distance

of complainant from call center, ease of communications, and responsiveness to call. To simplify the analysis, the customer satisfaction department began by analyzing only one variable: ease of communications. In order to analyze this variable, a scale from 1 to 5 was developed, with 1 meaning very difficult to communicate and 5 meaning very easy to communicate. Also, a customer satisfaction index was developed using the same scale, with a 1 meaning very unsatisfied and 5 meaning very satisfied. Ratings for each day were collected and averaged. Figure 9.7 shows the scatter plot for the past 100 days.

It can be seen that as ease of communications increases, the customer satisfaction index also increases; that is, there is a positive relationship between ease of communications and customer satisfaction index. Although there seems to be a positive relationship between these two variables, it does not seem to be too strong because the data points are too scattered around the regression line. So, a regression analysis was performed using statistical software. Figure 9.8 shows the results of this regression analysis.

The regression equation calculated by the analysis was

$$\text{Customer satisfaction index} = (1.403) + (0.639981) \times \text{Ease of communications}$$

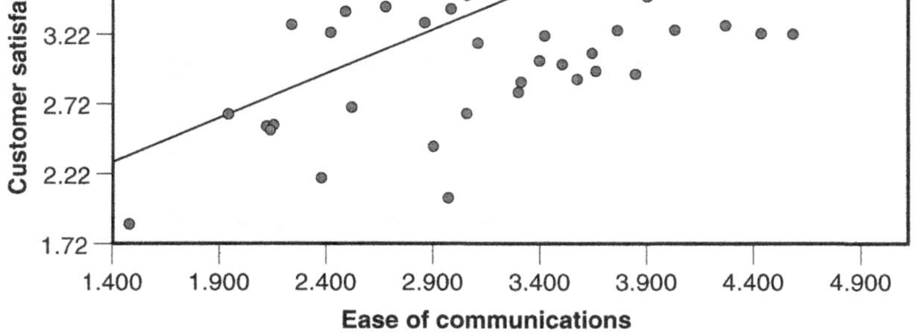

Figure 9.7 Scatter plot for ease of communications versus customer satisfaction index.

Simple regression model: Customer satisfaction index = (1.403) + (0.639981) × Ease of communications				
Model summary:				
R-squared	55.56%			
Parameter estimates:				
Predictor term	**Coefficient**	**SE coefficient**	**T**	**P**
Constant	1.403	0.222942	6.291	**0.0000**
Ease of communications	0.639981	0.057813498	11.070	**0.0000**

Figure 9.8 Regression analysis for ease of communications versus customer satisfaction index.

The slope (0.639981) demonstrates there is a positive relationship between ease of communications and customer satisfaction index. However, is that a strong relationship or a weak relationship? Taking a look at the R^2 (*R*-squared) value, it is 55.56%, or 0.5556. Taking the square root of that value gives us the value of the correlation coefficient, which is 0.745. Not a bad value. It seems there is a moderate relationship between ease of communications and customer satisfaction index. However, returning to the R^2 value, it is only 55.56%. What does that mean?

Recall from Section 9.3 that an R^2 value greater than 80% is considered acceptable. In this example, the obtained R^2 value means that only 55.56% of the regression model is explained by the simple linear regression equation; that is, there is another 44.44% of the model that is explained by some factors not considered in our regression model. So, let us turn our attention to multiple linear regression.

9.6 MULTIPLE LINEAR REGRESSION

As mentioned in Section 9.3, whenever the simple linear regression model provides a low value for the determination coefficient (for example, $R^2 <$ 80%), that means there are other variables not considered that are affecting the model. In such a case, we would need to identify which other variables could be affecting the model. The *multiple linear regression* model can be explained by the following formula:

$$y = \beta_0 + \beta_1 x_1 + \beta_2 x_2 + \beta_3 x_3 + \ldots + \beta_n x_n + \varepsilon$$

where y is the dependent variable, the x's (x_1, x_2, x_3, and so on) are the independent variables, β_0 is the y-intercept, and the other β's (β_1, β_2, β_3, and so on) are the slopes for each of the independent variables. There is also an error term, defined by ε. Let us explain the concept with an example:

> Since the medical device manufacturer in the previous example realized that other variables not considered in the simple regression model might be affecting the analysis, they decided to include other factors, such as distance from call center (expressed in miles), number of complaints received during the day, and responsiveness to calls. For the last one (responsiveness to calls) the same scale from 1 to 5 was used, 1 meaning poor responsiveness and 5 meaning excellent responsiveness. The results of the multiple linear regression analysis are presented in Figure 9.9.
>
> It can be noted that the R^2 value has increased from 55.56% to 90.10%. Now 90.10% of the model is explained by the following multiple regression equation:

$$\begin{gathered}\text{Customer satisfaction index} = \\ (0.408903) + (0.000293) \times \text{Distance from call center} + \\ (0.001892) \times \text{Number of complaints received} + \\ (0.433466) \times \text{Responsiveness to calls} + \\ (0.430775) \times \text{Ease of communications}\end{gathered}$$

Multiple regression model: Customer satisfaction index = (0.408903) + (0.000293) × Distance from call center + (0.001892) × Number of complaints received + (0.433466) × Responsiveness to calls + (0.430775) × Ease of communications

Model summary:

R-squared 90.10%

Parameter estimates:

Predictor term	**Coefficient**	**SE coefficient**	**T**	**P**
Constant	0.408903	0.232912	1.756	0.0824
Distance from call center	0.000293	0.002507031	0.116701	0.9073
Number of complaints received	0.001892	0.004489271	0.421513	0.6743
Responsiveness to calls	0.433466	0.02450545	17.689	**0.0000**
Ease of communications	0.430775	0.030548301	14.101	**0.0000**

Figure 9.9 Four-factor multiple regression analysis for customer satisfaction index.

But are all the factors significant for the regression equation? In Appendix D, the hypothesis test for regression analysis is presented. The null and alternate hypotheses are:

$$H_0\text{: Data are not correlated}$$

$$H_a\text{: Data are correlated}$$

Appendix D shows that when the obtained *p*-value is lower than an alpha value (α), the null hypothesis will be rejected and the alternate hypothesis will be accepted. So, in order to conclude that the input variable (x) is related to the output variable (y), we must obtain a *p*-value lower than the alpha value (α). Considering that $\alpha = 0.05$, all the input variables with a *p*-value lower than 0.05 will be considered significant; that is, those input variables are related to the output variables.

Taking another look at Figure 9.9, we can see that only *Responsiveness to calls* and *Ease of communications* have *p*-values lower than 0.05. We can conclude that those two input variables are related to the output variable, *Customer satisfaction index*. So, we will rule out the other two variables and run the analysis again. Figure 9.10 shows the results.

It can be seen that the R^2 value has only decreased from 90.10% to 90.08%. Now 90.08% of the model is explained by the following multiple regression equation:

$$\text{Customer satisfaction index} = (0.493463) + (0.435673) \times \text{Responsiveness to calls} + (0.433346) \times \text{Ease of communications}$$

Multiple regression model: Customer satisfaction index = 0.493463 + (0.435673) × Responsiveness to calls + (0.433346) × Ease of communications

Model summary:

R-squared 90.08%

Parameter estimates:

Predictor term	**Coefficient**	**SE coefficient**	**T**	**P**
Constant	0.493463	0.116857	4.223	0.0001
Responsiveness to calls	0.435673	0.023710993	18.374	**0.0000**
Ease of communications	0.433346	0.029667131	14.607	**0.0000**

Figure 9.10 Two-factor multiple regression alanysis for customer satisfaction index.

This regression equation can be used to predict the value of the customer satisfaction index based on the values of *Responsiveness to calls* and *Ease of communications*.

9.7 SUMMARY

The regression analysis provides useful information for the design of experiments, which will be explained in the next chapter. Regression can assist in the determination of which input variables have an impact on the output variables. It may also establish the kind of relationship between the input variables and output variables (positive correlation or negative correlation), and the magnitude of that relationship (strong or weak). The only drawback to regression is that it is done with already collected data. In that sense, there could be some uncontrollable factors affecting the results. For this reason, the design of experiments will be presented as a systematic tool for identifying the sources of process variability.

10
Design of Experiments

10.1 OVERVIEW

In industry, designed experiments can be used to systematically investigate the process or product variables that influence product quality. After you identify the process conditions and product components that influence product quality, you can direct improvement efforts to enhance a product's manufacturability, reliability, quality, and field performance. Because resources are limited, it is very important to get the most information from each experiment you perform. Well-designed experiments can produce significantly more information and often require fewer runs than haphazard or unplanned experiments. In addition, a well-designed experiment will ensure that you can evaluate the effects that you have identified as important.

The classical approach of changing one variable at a time has shortcomings:

- Too many experiments are necessary to study the effects of all the input factors.
- The optimum combination of all the variables may never be revealed.
- The interaction between factors (the behavior of one factor being dependent on the level of another factor) can not be determined

Classical experiments focus on OFAT (one factor at a time) at two or three levels and try to hold everything else constant ("blocked"), which is impossible to do in a complicated process. When a statistical designed experiment is properly constructed, it can focus on a wide range of key input factors or variables and will determine the optimum levels of each of the factors. Some of the benefits of statistical *design of experiments* (DOE) are:

- Many factors can be evaluated simultaneously.
- One can look at a process with relatively few experiments.
- If some (noise) factors can not be controlled, other input factors can be controlled.
- In-depth, statistical knowledge is not necessary to get a big benefit.
- Quality can be improved without cost increase.
- In many cases, tremendous cost savings can be achieved.

Factorial designs allow for the simultaneous study of the effects that several factors may have on a process. When performing an experiment, varying the levels of the factors simultaneously rather than one at a time is efficient in terms of time and cost, and allows for the study of interactions between factors. Interactions are the driving force in many processes. Without the use of factorial experiments, important interactions may remain undetected.

10.2 DESIGN OF EXPERIMENTS TERMINOLOGY

When dealing with DOE, some specific words come into play. Table 10.1 shows some of the most-used words in DOE, along with their counterparts in our day-to-day conversations about statistical tools.

So, a DOE can be defined as a systematic way to treat the factors at certain levels in order to evaluate the effect on the response variable. Or, in layman's terminology, to test the different combinations of inputs and settings in order to evaluate the result on the output variable.

In DOE there is a term called "noise." It refers specifically to an input that can not be controlled or is too difficult or costly to control. Some

Table 10.1 Design of experiments terminology.

Common term	Design of experiments term
Input (x)	Factor
Output (y)	Response
Setting	Level
Result	Effect
Uncontrollable input	Noise
Combination of inputs and settings	Treatment, run

examples might be humidity, raw material quality, operators, and so on. As will be explained later, in a *full factorial experiment*, all the factors are tested at all the levels. Traditionally, we select two levels for each factor (usually called "low" level and "high" level). This means that we need to specify certain settings that we can control. However, what do we do when we have a factor that can not be controlled or is too difficult or costly to control? There are many approaches to deal with this. However, one of the most commonly used approaches is "blocking." It essentially means to break our experiment into certain stages (called *blocks*) and run the same treatments on each block. Instead of adding another factor, we analyze the results on each of these blocks and determine if the block is significant or not. Blocking will be discussed in detail in Section 10.5.

10.3 FULL FACTORIAL EXPERIMENTS

As mentioned in Section 10.2, a *full factorial experiment* is one in which we evaluate all the possible combinations of factors and levels. That is, we run all the factors (inputs) at all levels (settings). The most common factorial experiment is called the 2^k, where 2 is the number of levels and k is the number of factors. In this type of experiment, if we have two factors, we will have four combinations; for three factors, we will have eight combinations; for four factors, we will have 16 combinations; and so on. The levels are usually identified as "low" and "high." These levels must be selected such that they are not too close together or too far apart. Too close is not good because we might not see a change in the response variable when one does exist; too far is not good because we might be experiencing unwanted nonlinear relationships.

But why do we not try with three or more levels to learn more about the process? For instance, set the levels to "low," "medium," and "high." In DOE, it is more desirable to run many small experiments than run too few big experiments. Table 10.2 shows the relationship between the number of factors and levels, and their impact on the number of runs for the experiment.

It can be seen in Table 10.2 that for each additional factor in a 2^k experiment, the number of combinations doubles; however, for each additional factor in a 3^k experiment, the number of combinations triples. Let us analyze the scenario where there are four factors. A 2^k experiment with four factors will require 16 experiments; however, a 3^k experiment with four factors will require 81 experiments. In theory, we could perform five of the 2^k experiments (80 runs) with fewer resources than a single 3^k experiment (81

Table 10.2 Relationship between number of levels and factors.

Number of levels	Number of factors	Number of runs
2	2	4
2	3	8
2	4	16
3	2	9
3	3	27
3	4	81

runs). So, which one do you think will provide the most information with the fewest number of experiments? The 2^k, of course.

10.4 FRACTIONAL FACTORIAL EXPERIMENTS

As mentioned earlier, in a full factorial experiment, we test all the possible combinations of levels and factors. It was also mentioned that as the number of factors increases, the number of experiments rises dramatically. So, what alternative do we have when the number of factors in an experiment is high (let's say, more than five factors)? In this case, we can perform what is called a *fractional factorial experiment.*

In contrast to a full factorial experiment, a fractional factorial experiment minimizes the number of runs. However, we can not fractionalize the experiment into any arbitrary number of runs. Since in a 2^k full factorial experiment the number of runs doubles with each additional factor, a similar approach will be followed for the fractional factorial. However, in this case, the number of runs will be half for each additional degree of fractionalization. For example, a five-level full factorial experiment will require 32 runs. However, we could run a half fractional factorial experiment with 16 runs, or a quarter fractional factorial experiment with eight runs, or an eighth fractional factorial experiment with four runs.

It is very important to realize that the degree of fractionalization can not be set arbitrarily. This is because it would be very tempting to run all fractional factorial experiments with the fewest number of runs possible (let's say, four runs). In fractional factorials there is something called the *resolution.* Technically, the resolution is a measure of the accuracy of the information provided by the experiment. The higher the resolution, the more accurate the information provided. The lower the resolution, the

Table 10.3 Degree of fractionalization versus resolution.

Number of runs	Number of factors													
	2	**3**	**4**	**5**	**6**	**7**	**8**	**9**	**10**	**11**	**12**	**13**	**14**	**15**
4	Full	III												
8		Full	IV	III	III	III								
16			Full	V	IV	IV	IV	III	III	III	III	III	III	III
32				Full	VI	IV	IV	IV	IV	IV	IV	IV	IV	IV
64					Full	VII	V	IV	IV	IV	IV	IV	IV	IV
128						Full	VIII	VI	V	V	IV	IV	IV	IV

Legend:
Red zone—Resolution III
Yellow zone—Resolution IV
Green zone—Resolution V and above

less accurate the information provided. That is, as resolution decreases, the amount of "confounded" effects increases. Confounding occurs when some factors (main factors or interaction factors) are literally aliased with other factors (interaction factors) so that the number of runs can be minimized and we are able to calculate the effects of the main factors. Table 10.3 shows the various resolutions for different degrees of fractionalization.

Let's say that our experiment has six factors at two levels. A full factorial experiment will require a total of 64 runs. However, we could perform a half fractional factorial experiment with 32 runs (resolution VI), a quarter fractional factorial experiment with 16 runs (resolution IV), or an eighth fractional factorial experiment with eight runs (resolution III). As can be seen in Table 10.3, a full factorial and a half fractional factorial experiment are good options because they are highlighted as *green zone* (resolution V and above). The quarter fractional factorial experiment is highlighted in *yellow zone* (resolution IV), so caution should be exercised because there might be some confounded effects. Finally, the eighth fractional factorial experiment is highlighted in *red zone* (resolution III), so it is not recommended to use this degree of fractionalization because there are too many confounded effects and the results might hide some important information.

10.5 BLOCKING

In the ideal DOE, all factors included in the experiments will be controlled. That is, all factors can be set to determined levels (settings). However, there are certain occasions when a factor can not be controlled or it could be too costly to control. For example, we would probably not be able to control the humidity level in a warehouse, or the variability in a supplied raw material, or the consistency among operators.

In each of these situations, we could try to include the variable as a factor in the experiment. However, how could you guarantee that warehouse humidity is exactly the same throughout the day? Or, how can you be certain that an operator has the same consistency throughout the day? These are only some examples of noncontrollable factors. So, how do we deal with those factors that can not be completely controlled or are too costly to control? An alternative would be to add "blocks" to our experiment.

A *block* is technically a set of conditions that will be replicated at certain times. Instead of adding a factor and analyzing that factor, we are going to add a block and determine if the block is significant or not. If the block is not significant, there aren't any major problems. This means that running our experiment with any of those conditions (let's say, humidity levels, supplied raw material, specific operator, and so on) will not have a significant impact on the response variable. However, if the block becomes significant, we need to determine how that factor (the block) will either be controlled or set to a fixed value. The first option (controlling the factor) might be the most difficult to achieve. So, sometimes when a block becomes significant, the best approach is to determine which of the levels provides the best results. Let's look at an example:

> During an experiment to determine the hardness of a tablet, a quality engineer found that raw material was an important factor to consider. The company is currently receiving that raw material from two different suppliers: supplier A and supplier B. Previous experience has shown that raw material is a noncontrollable factor (that is, raw material is considered a noise factor). Thus, it can not be included as a factor in the experiment because we are not able to control the levels (settings) of that factor. So, we will include the raw material as a block in our experiment. The other two important factors are *speed* and *compression force*. A two-level factorial experiment with two factors was generated. Table 10.4 shows the eight runs, along with the hardness result for each run. Notice the two blocks that were developed in this experiment: block 1 represents the raw material for supplier A, and block 2 represents the raw material for supplier B.
>
> Figure 10.1 shows the results when the experiment was analyzed using statistical software. It can be seen that the block is not significant because it has a high p-value (0.638). More on this will

Table 10.4 Data table for blocking experiment.

Blocks	A: Machine speed	B: Compression force	Hardness
1	1000	20	5
1	3000	20	8
1	1000	40	4
1	3000	40	10
2	1000	20	6
2	3000	20	9
2	1000	40	4
2	3000	40	9

Term	Coefficient	SE coefficient	T	P
Constant	6.750	0.339	19.941	**0.000**
A: Machine speed	2.125	0.239	8.878	**0.003**
B: Compression force	–0.125	0.239	–0.522	0.638
AB	0.625	0.239	2.611	0.080
Blocks—2	0.250	0.479	0.522	0.638

Figure 10.1 Results for blocking experiment.

be presented in section 10.8. From this experiment, we can conclude that raw material (that is, which supplier to use) does not have an impact on the tablet hardness. But, what if the *p*-value was very small (let's say, lower than 0.05)? In that case, raw material would be considered an important factor because, based on which supplier we do use, the tablet hardness would be different. Should we include this factor in our experiment? It depends. Remember that the levels in the experiment must be controllable. In the current situation, if raw material is not controllable, my opinion would be to use only raw material from the supplier that provides the best quality.

10.6 REPETITION AND REPLICATION

Very often when we are designing an experiment, we hear the words *repetition* and *replication*. They might seem like they are the same thing; however, each of these terms represents something different. Most people tend to use them interchangeably because the terms are somewhat confusing. Let me explain, in practical terms, what each of these terms represents.

When we are doing *repetitions* in DOE, we are actually not repeating anything. What we are doing is just taking more than one sample from each run. That is, for each combination of factor and level, we obtain more than one result. What is the advantage of this? With more than one datum, we can calculate descriptive statistics such as average, median, range, standard deviation, and so on. Out of these descriptive statistics, the measures of dispersion (range, standard deviation, and variance) play an important role. Specifically, from each run we can calculate what is called *short-term variability*. The logic is, if we are keeping things constant within a run, why would we see variability in the results? The answer is, because the variation is inherent in the process itself *at that moment*. Table 10.5 shows a *repeated* experiment.

Notice that for each run, two samples were taken. We could use that information to calculate the measures of central tendency (average, median, and so on) and the measures of dispersion (range, standard deviation, and so on). Then, our experiment would not only be focused on hitting a target (for instance, the nominal value of the specification), but also on reducing the variation (that is, obtaining the optimal combination of factor and level that provides the smallest variation possible).

In contrast, when we change the conditions between each run, we will be able to calculate the *long-term variability*. When we do this, we are using *replications* within our experiment. In a replicated experiment, we are capturing the process variation in the long run. That is, we are also considering the effect of some factors that were not included in our experiment. What is the logic? If we run a certain combination of factor and level at some point, we might expect that the same combination could be run at any other time and the result would be the same or very similar. If that presumption does not hold true, then there might be some other factors not considered that are influencing the response variable. In that case, we would need

Table 10.5 Repetition in DOE.

A: Machine speed	B: Compression force	Hardness 1	Hardness 2
1000	20	5	5
3000	20	8	7
1000	40	4	4
3000	40	10	11

to extend our experiment to include such factors. Table 10.6 shows an example of a *replicated* experiment.

So, what is the best approach? To consider *both repetition and replication* in our experiment. Table 10.7 shows an example where we consider both.

Notice that *repetition* can be observed for each run. Two tablets were collected for each combination of factor and level. As mentioned, short-term variability can be calculated from this approach. *Replication* can be observed for run #1 and #5 (1000, 20), run #2 and #6 (3000, 20), run #3 and #7 (1000, 40), and run #4 and #8 (3000, 40). Long-term variability can then be calculated.

Table 10.6 Replication in DOE.

A: Machine speed	B: Compression force	Hardness 1
1000	20	5
3000	20	8
1000	40	4
3000	40	10
1000	20	6
3000	20	9
1000	40	4
3000	40	9

Table 10.7 Repetition and replication in DOE.

A: Machine speed	B: Compression force	Hardness 1	Hardness 2
1000	20	5	5
3000	20	8	7
1000	40	4	4
3000	40	10	11
1000	20	6	5
3000	20	9	9
1000	40	4	4
3000	40	9	10

But what is the practical application of considering short-term and long-term variability in our experiments? One of the most common errors in experimental design is to focus just on hitting the target regardless of the variation. The best experiment would consider both. Let us imagine that the nominal value is 10 kp for the average tablet hardness, with a tolerance from 6 to 14 kp. Suppose that one combination of factor and level resulted in an average hardness of 10 kp and a standard deviation of 1.0 kp. This means that 99.73% of the data will fall between 7 and 13 kp. However, there was another combination that resulted in an average hardness of 11 kp and a standard deviation of 0.5 kp. This would mean that 99.73% of the data will fall between 9.5 and 12.5 kp. Which combination of factor and level would provide the more consistent product? The second combination (9.5 to 12.5 kp), of course. However, if we had just focused on the central tendency, we would have chosen the first combination.

Another way to see the impact of just focusing on central tendency is to calculate the capability indices for both scenarios. For the first combination (average = 10 kp, standard deviation = 1.0 kp), C_p will be 1.33 and C_{pk} also 1.33. This is not so bad, as we already mentioned in Section 7.3. But what would be the capability indices for the second combination (average = 11 kp, standard deviation = 0.5 kp)? The answer is $C_p = 2.67$ and $C_{pk} = 2.00$. Again, which combination provides the best result? The one that is somewhat off-target but with the smallest variation. In summary, the next time you perform a DOE, remember to consider both descriptive statistics measures: central tendency and dispersion.

10.7 EXPERIMENTAL STRATEGY

In Sections 10.3 and 10.4, I presented the concepts of full factorial and fractional factorial experiments, respectively. The differences between these two types of experiments were explained in those sections. Advantages and limitations of each one were also explained. But when is it more appropriate to use each one of these types of experiments? The answer is related to the amount of knowledge we have about the process we want to study.

When the process knowledge is low (because it is a new process or it has not been thoroughly studied), the best approach is to start with a fractional factorial experiment. In this way, we can experiment with many variables and learn more about their effect on the response variable(s). Just remember that experimentation is sequential; that is, we will apply the knowledge acquired in each experiment to subsequent experiments. Once our process knowledge becomes greater, we can then start using full factorial experiments. But, remember, one limitation with full factorial experiments is that the number of experiments becomes larger as the number of factors increases. So, our first full factorial experiments must not be replicated. At this point, we still want to keep the number of runs as low as possible.

Once we are certain we have included the appropriate factors in our experiment (at the appropriate levels), we would like to add replications to our experiments. In this way, long-term variability will be accounted for. A few full factorial *replicated* experiments will help us make certain we have considered all the significant factors and levels. Finally, a few validation runs would be appropriate once we have found the factors and levels that optimize our response variable(s).

Following the approach outlined above, we can be confident that we are using the available resources in the most efficient way and we are continuosuly expanding our process knowledge.

10.8 DESIGN OF EXPERIMENTS EXAMPLE: TWO LEVELS, TWO FACTORS

Let me explain the concept of a two-level, two-factor full factorial experiment with an example:

> A company has been performing some experiments in order to optimize the hardness of a tablet. The specification for the tablet hardness is 10 ± 0.5 kp. So far, the company has acquired much process knowledge through previous experimentation. During this process, the company has performed fractional factorial and unreplicated full factorial experiments. The company has also experimented with center points and blocks to rule out curvature and noise factors, respectively. At this point, a replicated full factorial experiment will be performed.
>
> The two factors that have been demonstrated to be more significant throughout all the experimentation stages in the compression process are *machine speed* and *compression force*. The levels to test in this experiment will be 1000 and 3000 rpm for the machine speed and 20 and 40 kn for the compression force. A replicated two-level, two-factor experiment was developed. The results are presented in Table 10.8.
>
> Using statistical software, the results were analyzed. Figure 10.2 shows the main effects plots for this experiment. How is the main effects plot analyzed? In the *x*-axis, you will find the two levels for each factor (1000 and 3000 rpm for machine speed, and

Table 10.8 Replicated full factorial design example for tablet hardness.

A: Machine speed	B: Compression force	Tablet hardness
3000	20	8.8
1000	20	9.2
3000	20	8.1
1000	40	9.1
3000	40	11.6
1000	20	8.7
1000	40	8.3
3000	40	11.0

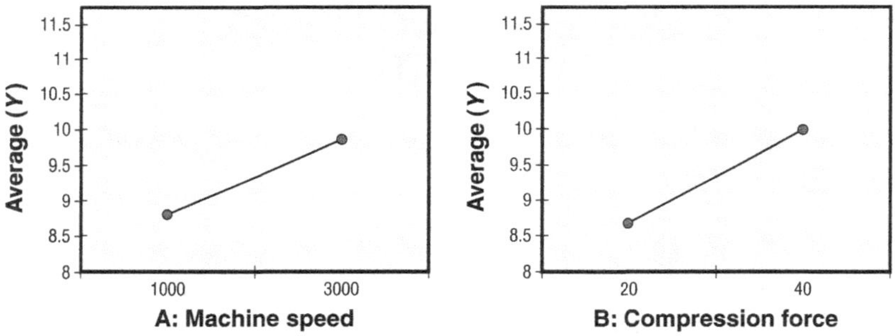

Figure 10.2 Main effects plots for tablet hardness example.

20 and 40 kn for compression force). In the *y*-axis, you will find the average tablet hardness at each level. The steeper the line, the more significant is that factor. When the line is completely flat, it does not matter which level you choose; the average response would be the same. In this example, in order to achieve an average hardness of about 10.0 kp, the machine speed should be set at 3000 rpm and the compression force at 40 kn.

But is the interaction between the machine speed and compression force significant? That is, would selecting a different machine speed or compression force cause a difference in the response variable? Figure 10.3 shows the interaction plots for the experiment.

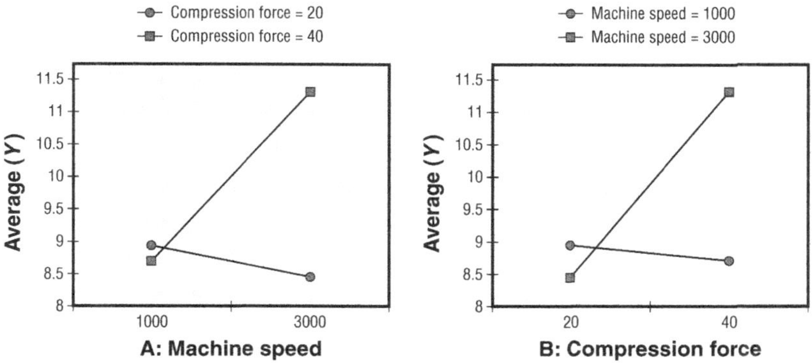

Figure 10.3 Interaction plots for tablet hardness example.

How will we determine the strength of the interaction between machine speed and compression force? If the lines are completely perpendicular, there is a strong interaction; if the lines are completely parallel, there is no interaction. In this example, the lines are neither completely perpendicular nor completely parallel. However, the lines are closer to perpendicular than parallel; so, we can conclude that there is a strong interaction between machine speed and compression force.

As mentioned earlier when analyzing the main effects plots, by setting the machine speed to 3000 rpm and the compression force to 40 kn, we achieve an average hardness of about 10 kp, which is the nominal value. But what if, leaving machine speed at 3000 rpm, we decide to set compression force to 20 kn? The plot on the left side of Figure 10.3 shows that average hardness would drop significantly. The same will happen if we set the compression force at 40 kn but set the machine speed at 1000 rpm, as shown on the right side of Figure 10.3.

So, we can conclude that both the main effect factors and the interaction factors seem to be significant, based on these plots. However, as mentioned earlier, we can not rely only on the graphs to make a conclusion about the significance of the main effects and interaction factors. So, we need to study the analytical portion of the experiment. Figure 10.4 shows the results obtained using statistical software.

Based on the results in Figure 10.4, at an alpha level of 0.05 (α = 0.05), we can conclude that both main effects (machine speed and compression force) and the interaction of those effects

Term	Coefficient	T	P
Constant	9.350	56.706	**0.0000**
A: Machine speed	0.525	3.184	**0.0334**
B: Compression force	0.650	3.942	**0.0169**
AB	0.775	4.700	**0.0093**

Figure 10.4 Factorial design analysis for tablet hardness example.

are significant. More information about when to reject and fail to reject H_0 is provided in Appendix D for the most commonly used hypothesis tests.

10.9 SUMMARY

So far, we have been analyzing data that were previously collected. From that data, different evaluations have been perfomed: practical, graphical, and analytical. The tools studied so far have assisted us to get a better understanding of our processes. Now we want to use that knowledge to find the inputs and settings that will optimize the results, through an approach called design of experiments (DOE).

The type of DOE to perform will depend on the process knowledge. When process knowledge is low, we will start experimenting with fractional factorials. Then, as process knowledge becomes greater, we will experiment with full factorials. When using fractional factorials, the level of fractionalization is an important element to consider. We will select the level of fractionalization based on the resolution provided by each alternative. Resolution III must never be run; resolution IV experiments must be used with caution; resolution V and above are recommended.

When performing full factorial experiments, we could start with unreplicated experiments. Then, as process knowledge becomes greater, we could add replicates in order to calculate the long-term variability. Also, repetitions could be added in order to calculate the short-term variability. Experimentation must not focus only on achieving a target, but also on reducing variation.

11

Control Charts

11.1 OVERVIEW

In Chapter 3 the concept of process variation was introduced. The assumption is that all processes are subject to some kind of variation. Two types of variation were defined: *common cause* variation and *special cause* variation. As mentioned in that chapter, common cause variation is always present in every process because no process is perfect. Common cause variation is inherent in every process. In contrast, special cause variation is not always present in every process. This type of variation is caused by assignable events, that is, by certain things that have a significant impact on the process.

In Chapter 5 some graphical tools for analyzing data were presented. Tools like the histogram, box plot, dot plot, and Pareto diagram (among others) were presented and discussed in that chapter. The major disadvantage of those tools is that they represent the data in a static way. For instance, a histogram can help us describe the data in terms of central tendency, dispersion, and shape. However, the histogram does not tell us anything about each individual value in terms of *time*. How could we obtain the advantages of learning about central tendency, dispersion, and shape along with the advantages of learning how the process behaves over time? Simply by using a *control chart.*

Recall from Chapter 7 that there is something we call the *process spread*, also known as the *voice of the process*. The process spread is quantified by the standard deviation, σ. Specifically, it is defined by the interval of $\pm 3\sigma$ from the mean. As mentioned in Section 4.5, in a normal distribution about 99.73% of the data is expected to fall within $\pm 3\sigma$ from the mean. So, we will use that fact in order to calculate what are called *control limits*. In a control chart, the control limits define where the common causes of variation are expected to lie. That is, as long as the process is in statistical control, all the points will lie within the control limits defined by the interval of $\pm 3\sigma$ from the mean, without any nonrandom pattern, as will be studied later. When we see a point outside of those control limits (or points showing a nonrandom pattern), that indicates some sort of assignable or special cause that needs to be studied and corrected.

11.2 THE RATIONAL SUBGROUP

One of the most important concepts in control charting is the *rational subgroup*. The success of the control chart will depend, in part, on the appropriate selection of the subgroup size. A rational subgroup is just a subset of a group intended to be as homogeneous as possible. In that way, we will be able to compare the variation within each subgroup and compare the variation between subgroups over an extended period. But how do we determine the appropriate subgroup size? That will depend on the process being studied.

Historically, many people use a subgroup size of 5 because it is very convenient for most manufacturing processes. However, a subgroup size of 5 is not always representative of the process. For instance, let us imagine that we are studying a molding process. The mold has 10 cavities. In this case, it would be preferable to select a subgroup size of 10 instead of a subgroup size of 5. Why? Because the subgroup size of 10 will represent the variation of all the parts produced by that mold at a specific time. In that way, a control chart could be developed for that process and take, for instance, the 10 parts produced by the mold at a certain moment every hour. We could then compare the variation within each hour with the variation between hours to see if there are any significant differences in the long run.

11.3 NONRANDOM PATTERNS

As mentioned earlier, if a process is in statistical control, all the points will lie within the control limits without showing any nonrandom pattern. But what are those nonrandom patterns we need to look for? Walter Shewhart developed a list of eight nonrandom patterns that might show that something is changing (or has changed) in the process. Every statistical software package has these eight tests, which may be applied individually or in any combination. Figure 11.1 shows the tests for nonrandom patterns to look for when analyzing control charts.

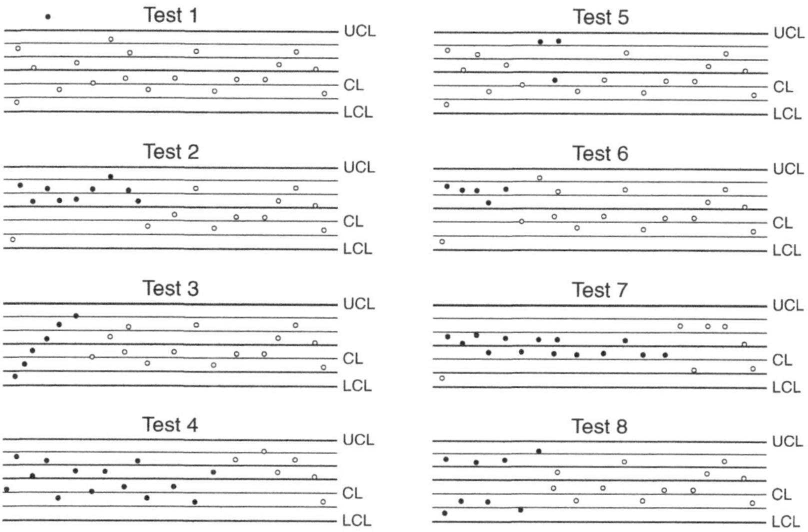

Figure 11.1 Tests for nonrandom patterns in control charting.

Do we need to adjust our process every time we see one of these eight nonrandom patterns? Not necessarily. Recall from Chapter 7 that being out of control does not necessarily mean producing defective parts. If a process is capable (that is, the process variation is narrower than the customer specifications), and if a point falls outside of the control limits or the chart starts to show a nonrandom pattern, we might still have time to react before the process produces nonconforming parts. So, let me explain some of the possible causes of each of the eight nonrandom patterns in order to provide an idea of what steps could be taken to adjust the process before it is too late:

- *Test 1—One point more than 3 sigma from centerline.* This is typically what is called a *special cause.* It is a point produced by an assignable event, some sort of change that caused an extreme variation in the process. Some examples are change of material, change of supplier, change in methods, and so on. Using a fishbone diagram could help us identify the potential causes of this kind of situation.
- *Test 2—Nine points in a row on same side of centerline.* This is usually the result of a change in the process centering. Although

the process variation might have remained constant, the process has shifted toward one of the control limits. This type of pattern does not necessarily mean that something bad has happened. For instance, we might see a reduction in the average time to complete an investigation after a CAPA certification process if we see this kind of pattern. In contrast, we might see this pattern on a filling process if one or various nozzles become clogged. In any case, this is an indicator that something has happened, and it is a good time to analyze the desirability of such change.

- *Test 3—Six points in a row, all increasing or all decreasing.* This might be an indication of tool wear, machine deterioration, tired operator, and so on. It does not represent a sudden change in the process, but a slight and continuous change in it. This kind of pattern can be easily detected and acted on before it is too late.
- *Test 4—Fourteen points in a row, alternating up and down.* This is an uncommon pattern. This can be caused by overadjustment of the equipment or by manipulation of data. Special attention must be given to data integrity.
- *Test 5—Two out of three points in a row more than 2 sigma from the centerline (same side).* This pattern might be indicative of a sudden increase in the process variation. It is possible to have some points in this zone from time to time, but two out of three consecutive points is not desirable. The causes of this pattern might be similar to those of test 1; however, in this case the the event has not been so significant as to cause an out-of-control point.
- *Test 6—Four out of five points in a row more than 1 sigma from the centerline (same side).* This might be the beginning of the pattern depicted in test 5. Instead of a sudden increase in variation as in test 5, here we have a slight increase in variation that, if not acted on, can result in out-of-control points.
- *Test 7—Fifteen points in a row within 1 sigma from the centerline (either side).* This pattern is indicative that variation has been dramatically reduced. This might look like "too good to be true" or "do not touch the process." The main problem here is that a variation reduction has been achieved, but the control limits have not been recalculated. Although a pattern like this might look good, it is not statistically correct. Remember that control limits are based on process variability. So, if variability decreases, then the control limits must be recalculated and become narrower.

- *Test 8—Eight points in a row more than 1 sigma from the centerline (either side).* This is a very interesting pattern. It might be indicative of mixtures, that is, combining data from different processes in the same control chart. An example could be to have data from two different machines (each one with a different average) plotted on the same control chart. The solution could be to separate the data for each machine into different control charts.

The situations presented above are not intended to cover all the possible causes of variation; they are just some examples of what could be causing each of these nonrandom patterns. The person analyzing the control charts must study the process and find the real causes of such variation. It is also important to note that the previous eight rules apply to the variables control charts as they will be defined in Section 11.4. However, in the attributes control charts, only test 1 through test 4 will be applicable. That means test 5 through test 8 will not apply to the attributes control charts. So, if you see one of these patterns in an attributes control chart, they will not be considered nonrandom patterns.

11.4 VARIABLES CONTROL CHARTS AND ATTRIBUTES CONTROL CHARTS

In Section 4.2, the different types of data were presented: variable data, attribute data, and locational data. Out of those three types of data, the most used are the variable and attribute data. As mentioned in that section, variable data is *continuous*—data that can be measured. In contrast, attribute data is *discrete*—data that can be counted, categorized, binary, and so on. Depending on the type of data at hand, we could use a different control chart.

One of the most common errors I have seen is the selection of the incorrect control chart. For instance, let's say we want to plot the number of complaints received during each month. Since the number of complaints is data that can be counted (discrete), we must use an attributes control chart, in this case, a *c*-chart. Many times, I have seen the use of a variables control chart (like an individuals and moving range chart) to plot this type of information. Other common errors I have seen in the use of control charts are:

- Wrong formula used to calculate control limits
- Missing, poor, or erroneous measurements

- Data on charts not current
- Process adjustments have not been noted
- Control limits and average not updated
- Special-cause signals ignored
- Nonrandom patterns not studied
- Specification limits placed on chart instead of control limits

Let me present the different types of variables and attributes control charts available, with some applications for each one.

11.5 VARIABLES CONTROL CHARTS

As mentioned, the variables control charts will be used for continuous data, or data that can be measured. Most parameters in a manufacturing process fit this type of data. The most commonly used variables control charts are the individuals and moving range (ImR) chart, $\bar{X}$ and R chart, and $\bar{X}$ and s chart. Once we determine that our data are continuous, we need to decide which of these charts is the most appropriate. So, how do we determine which variables control chart to use? It will depend on the subgroup size, as mentioned in Section 11.2.

So, here are some hints about which chart to use. If the subgroup size is 1, then we will use the ImR chart; if the subgroup size ranges from 2 to 5, then use an $\bar{X}$ and R chart; if the subgroup size is greater than or equal to 6, then use an $\bar{X}$ and s chart. Recall from Section 11.2 what is defined as a subgroup and what should be the appropriate subgroup size for each process.

11.5.1 Individuals and Moving Range Chart

Let us suppose that we want to plot the pH of a sample for a certain process. Each individual sample is collected in an individual bottle. Since the pH within the bottle will be the same (the sample is homogeneous), it does not make any sense to calculate the average of the sample at different locations in the bottle. Instead, one sample will be taken from each bottle and plotted in an ImR chart. Figure 11.2 shows the data for 100 consecutive bottles.

The specification for this process is 7.0 ± 1.0. Although all the points are well within the specification limits, there are three points in the moving range chart that fall outside of the upper control limit. When each individual sample was analyzed, it was found that those out-of-control points were

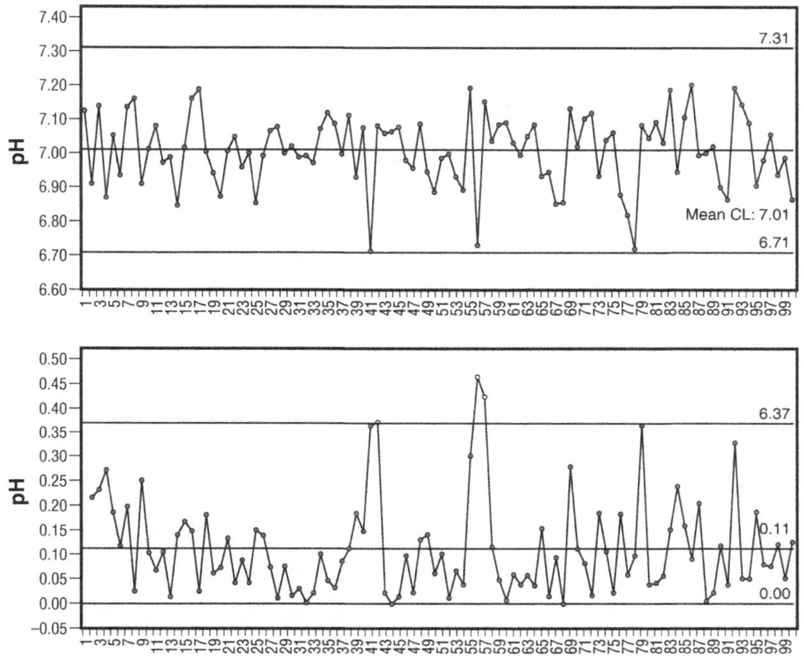

Figure 11.2 Individuals and moving range chart for pH.

the result of a sudden drop in sample #41 (from 7.08 to 6.71) and sample #56 (from 7.19 to 6.73). However, since those individual values are well within the specification limits and those sudden drops did not happen again, it was decided that no changes in the process are required. A common error would have been to overreact to those individual values. Overreacting would have caused more points to lie outside of the control limits on both charts.

11.5.2 $\bar{X}$ and *R* Chart

As mentioned earlier, when the subgroup size varies from 2 to 5, it is recommended to use the $\bar{X}$ and *R* chart. This is one of the easiest charts to use because both metrics are very simple to calculate. What is the logic behind limiting the use of this chart to subgroup sizes from 2 to 5? Recall that in order to calculate a range, only two values are needed: the highest value and the lowest value. So, supposing that each value is different, here are the different scenarios for calculating the range:

- For a subgroup size of 2, both values will be used.
- For a subgroup size of 3, two values will be used and one value will be discarded.
- For a subgroup size of 4, two values will be used and two values will be discarded.
- For a subgroup size of 5, two values will be used and three values will be discarded.

For subgroup sizes of 6 and above, the range is not a good measure of process dispersion because too many values will be discarded in the calculation. Instead, for such subgroup sizes, the standard deviation (or variance) is recommended because it uses all the individual values. Figure 11.3 shows the $\overline{X}$ and R chart for the tablet weight process of a certain company. Each point in the $\overline{X}$ chart represents the average of the weight of five tablets taken at a certain time, while each point in the range chart represents the range of those five tablets.

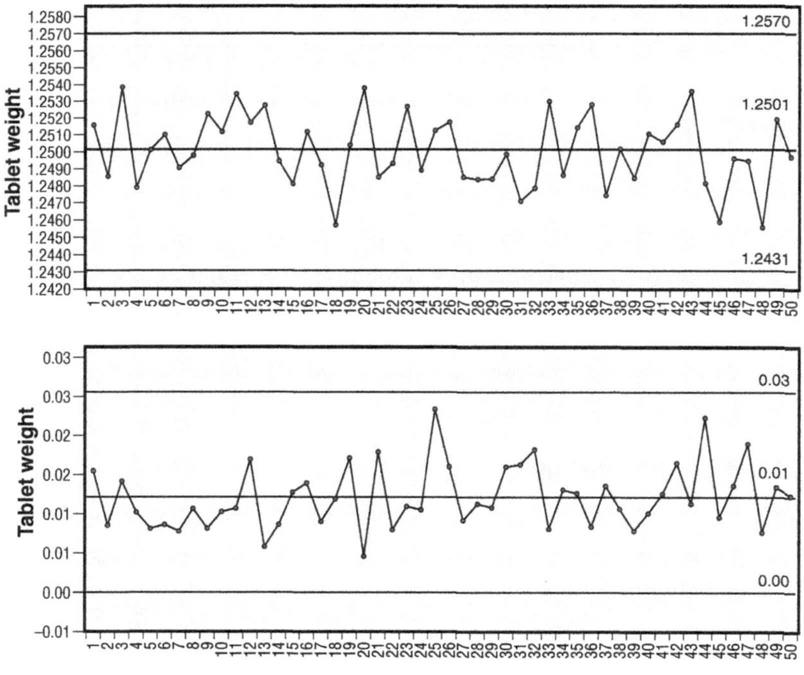

Figure 11.3 $\overline{X}$ and R chart for tablet weight.

It can be seen that the process is in statistical control, without any nonrandom pattern.

11.5.3 $\bar{X}$ and s Chart

Finally, when the subgroup size is 6 and above, it is recommended to use the $\bar{X}$ and s chart. As mentioned, the range is an easy metric to calculate because it is the difference between the largest and the smallest value. However, the standard deviation is somewhat more difficult to calculate. It requires the use of a statistical calculator or a spreadsheet. For that reason, it is seldom used. However, remember that ease of use must not be the main consideration in deciding which chart to use; the main reason must be subgroup size. As subgroup size increases, we are inclined to use the $\bar{X}$ and s chart over the $\bar{X}$ and R chart. Figure 11.4 shows the $\bar{X}$ and s chart for a bottle weight process. The machine fills 10 bottles at a time; that is, it has 10 nozzles. Instead of considering each individual bottle's weight, the

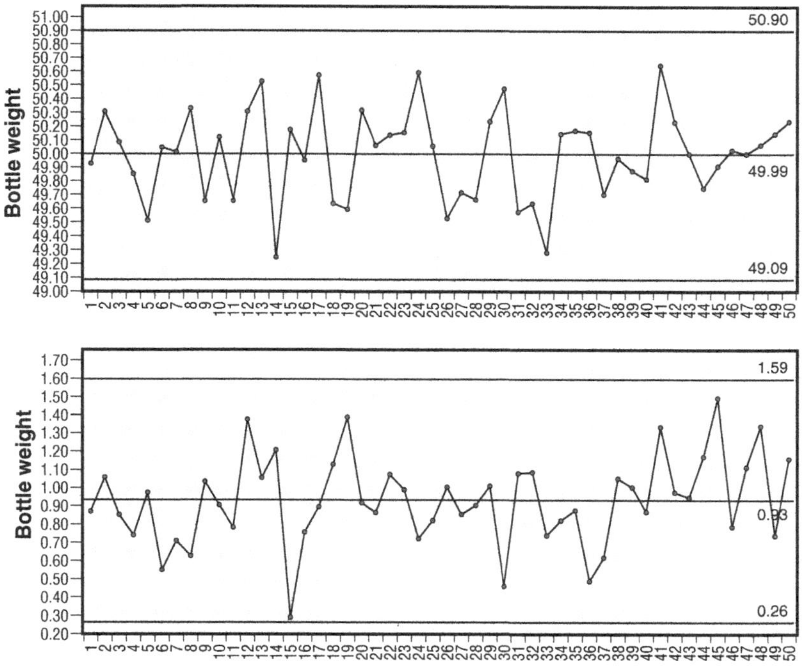

Figure 11.4 $\bar{X}$ and s chart for bottle weight.

company decided to monitor the average weight at certain specific times. So, an $\bar{X}$ and s chart was developed.

It can be seen that the process is in statistical control, without any nonrandom pattern.

11.6 ATTRIBUTES CONTROL CHARTS

If data are discrete, the variables control charts mentioned earlier can not be used. Instead, we need to use the attributes control charts. The most commonly used attributes control charts are the *p*-, *np*-, *c*-, and *u*-charts. So, how do we determine which attributes control chart to use? It will depend on what we want to plot: defectives or defects. A *defective* unit is a unit that has at least one defect. In contrast, a *defect* is any characteristic that does not conform to the specifications. As opposed to the variables control charts, in which two charts are usually plotted on the same page (one chart for central tendency and another for dispersion), in the attributes control charts we only plot one chart at a time. So, what is the difference between the different types of attributes control charts?

The *p*-chart and *np*-chart are used for *defectives*. Specifically, the *p*-chart is used to plot the *percentage defective*, while the *np*-chart plots the *number of defectives*. On the other hand, the *c*-chart and *u*-chart are used for *defects*. Particularly, the *c*-chart is for *number of defects*, while the *u*-chart is used for *average defects per unit*. I will use Figure 11.5 to

Figure 11.5 Attributes chart example.

	p-chart (% defectives)	*np*-chart Number of defectives	*c*-chart Number of defects	*u*-chart Average defects per unit
Subgroup 1	2/3 = 0.67 = 67%	2	3	3/3 = 1.00
Subgroup 2	1/3 = 0.33 = 33%	1	1	1/3 = 0.33
Subgroup 3	3/3 = 1.00 = 100%	3	4	4/3 = 1.33

Figure 11.6 Calculations for attributes charts example.

illustrate the difference between the different types of charts using the same data.

Each cylinder represents a single part. There are three subgroups, each of size 3. The "X" symbol within the cylinder represents a single defect. Figure 11.6 shows the calculated values for each of the four types of attributes control charts for each of the three subgroups.

It can be seen that we can analyze the same data in different ways by using different control charts. It just depends on which information we want to present.

11.6.1 *p*-Chart

The *p*-chart is used to analyze the percentage defective in each subgroup. As stated, it does not consider how many defects a unit might have; it considers the unit as defective if it has at least one defect. One important consideration for the *p*-chart (and for the *u*-chart, as will be presented later) is that subgroup size does not have to remain constant. The reason is that what we are plotting is the percentage of defectives, regardless of what the sample size of each subgroup is. For example, if we have one defective in a sample of three units, it will result in 0.33, or 33%, percent defective. On the other hand, if we have three defectives in a sample of nine units, it will also result in 0.33, or 33%, percent defective. Let me explain the use of a *p*-chart with an example:

> A company is performing an audit of their manufacturing batch records. They want to analyze what percentage of the batch records have any type of error, regardless of the number of errors or type of error. The reason is that any amount or type of error in the manufacturing batch record will render the product adulterated, as established in 21 USC §501. That is, the manufacturing batch record will be considered defective when it has at least one defect. Figure 11.7 shows the information for the records produced during the past year.
>
> As can be seen, the average percentage of manufacturing batch records with errors is 0.02, or 2%. The upper control limit is 5% and the lower control limit is 0%. As long as all the data points are within those limits, there are only common causes acting on the process. When a data point falls above the upper control limit (5%, in this example), then special causes are acting on the process. Please note that being within the control limits (between 0% and 5%, in this example) only means that the process is in statistical control. It does not mean that it is acceptable. For this process variable (percentage of manufacturing batch records with errors), the goal must be zero. This control chart could be the baseline for a process improvement project about manufacturing batch records error elimination.

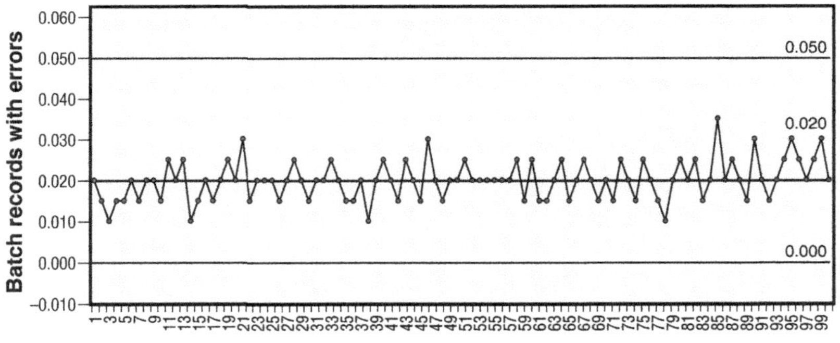

Figure 11.7 *p*-chart for percentage of manufacturing batch records with errors.

11.6.2 *np*-Chart

As mentioned, the *p*-chart is used to plot the percentage of defectives. Sometimes, we do not want to plot the percentage of defectives but the number of defectives. The reason? Mathematically, three defectives out of 10 units is 30%; but 300 defectives out of 1000 units is also 30%. When dealing with devices such as pacemakers, for example, it is more important that you analyze number of defectives than percentage of defectives, especially if you are one of the people receiving the pacemaker. One disadvantage of using an *np*-chart instead of a *p*-chart is that for the *np*-chart the subgroup size must remain constant. If the subgroup size varies, and we want to plot defectives, then the *p*-chart must be used. Let us use the same data as for the example of percentage of manufacturing batch records with errors, but this time focus on the number of defectives instead of the percentage defective. Figure 11.8 shows the results for the records produced during the past year.

Recall from the previous example that the control limits just establish the boundaries within which the process is considered to be in statistical control; they do not represent what is considered acceptable. In this example, having fewer than 9.86 defective units means that the process is in control. Again, as mentioned earlier, this control chart could be the baseline for a process improvement project about manufacturing batch records error elimination.

11.6.3 *c*-Chart

In Sections 11.6.1 and 11.6.2, I presented the attributes control charts for defectives. Recall that a defective unit is a unit that has at least one defect.

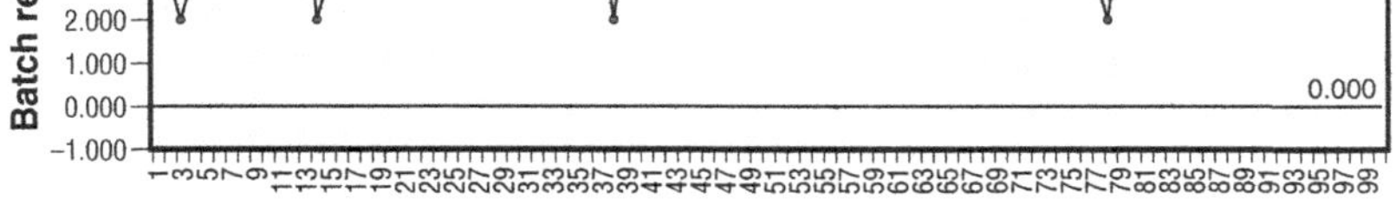

Figure 11.8 *np*-chart for number of manufacturing batch records with errors.

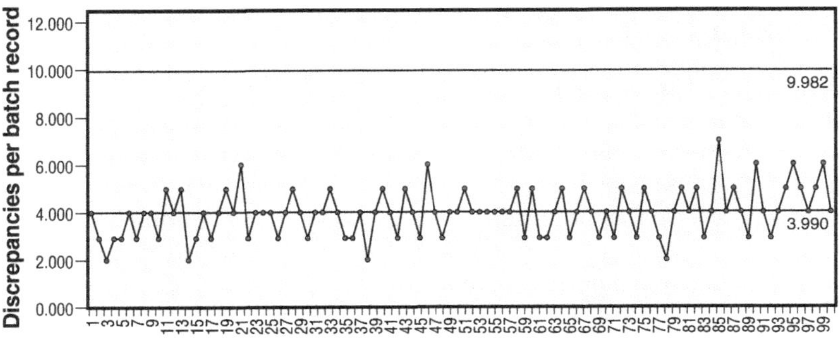

Figure 11.9 c-chart for number of errors per manufacturing batch record.

Monitoring the *defective* units might be good sometimes, but having knowledge about the *defects* is also important. The *c*-chart is used to plot the number of defects in each subgroup. As with the *np*-chart, one disadvantage of the *c*-chart is that the subgroup size must remain constant. Let us illustrate the application of the *c*-chart with an example:

> A company is performing an evaluation of the number of errors found in the manufacturing batch records. Although just one error makes the batch record defective, the team wants to analyze the number of errors in order to see the magnitude of the problem and start a project geared toward elimination of errors in the manufacturing batch records. Figure 11.9 shows the *c*-chart for number of errors in each batch record. The last 100 batch records were analyzed and plotted in sequential order.
>
> It can be seen that as long as the number of errors in each batch record remains between zero and 9.98, the process is in control. But, recall from the previous control charts that having errors in the manufacturing batch records is unacceptable. Thus, this control chart can be used to set the baseline and monitor the improvement. The goal is to eventually eliminate the errors in the manufacturing batch records.

11.6.4 *u*-Chart

As mentioned, one disadvantage of the *c*-chart is that subgroup size must remain constant. But what if the sample size in each subgroup is different? In that case, if we want to plot information about defects, we need to use the *u*-chart instead of the *c*-chart. Let us suppose that the company produces

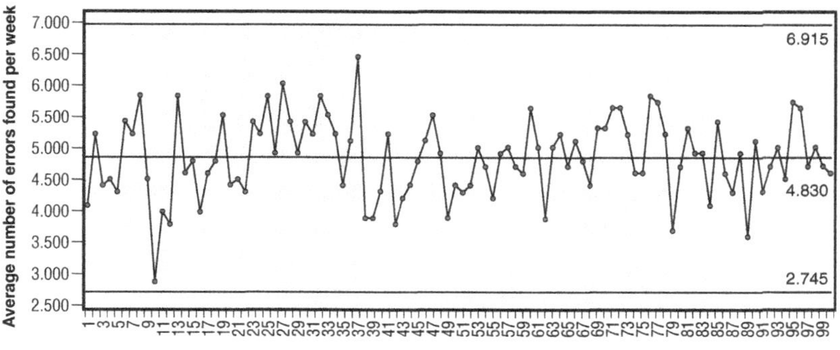

Figure 11.10 *u*-chart for average number of errors per batch record per week.

so many records that it is almost impossible to analyze all of them. So, they decided to take a sample of all records generated each week and plot the data for the average number of errors found each week. A subgroup of 10 records was collected each week. The results are presented in Figure 11.10.

Each data point represents the *average* number of errors per batch record each week. Some records might have more than the average, while others might have less than the average. As mentioned earlier, in the FDA-regulated industry, the target must be zero defects. So, this control chart should be used to set a baseline to monitor the improvement over time.

11.7 SUMMARY

Most of the graphical tools studied so far focus on looking at the central tendency, dispersion, and shape of the distribution. They also focus on making comparisons between various groups. However, none of them consider *when* each of the data points were collected. Time is an important consideration in every process. So, plotting the data as they are being collected can assist us in taking action before any major problem arises.

In previous chapters, we presented different types of data: attribute, variable, and locational. So, when plotting data in a control chart, one of the first issues to consider is which chart to use. For variable data, some of the most commonly used charts are the individuals and moving range (ImR) chart, $\overline{X}$ and R chart, and $\overline{X}$ and s chart. The criterion for selecting the apropriate variables chart will be the subgroup size. On the other hand, for attribute data, the most commonly used charts are the *p*-, *np*-, *c*-, and *u*-charts. The criterion for selecting the appropriate attributes chart will be whether we want to plot data for *defectives* or for *defects.*

When analyzing control charts, one of the aspects we will consider is the randomness of the data. There are eight rules for determining whether the process is exhibiting a random pattern or if there are special causes acting on the process. As long as all the data points are within the control limits, without any nonrandom pattern, we can say that the process is in statistical control; said differently, only common causes of variation are present. However, when there are data points outside of the control limits, or some nonrandom pattern is seen, we say that the process is out of statistical control. In this case, we have a combination of common causes and special causes of variation present in the process. Special causes must be identified and eliminated, while common causes can be reduced.

Statistical process control is a requirement for all the tests discussed in this book. So, although control charting was the last tool discussed, it will be one of the first tools to be used prior to any other analysis.

12
Final Thoughts

12.1 OVERVIEW

Throughout the preceding chapters, I have presented several tools to achieve process control in an organization regulated by the Food and Drug Administration (FDA). The book started by establishing the regulatory importance of statistical process control. The differences among regulations, guidances, and international standards were established. Then, some examples of the use (and misuse) of statistical tools, as evidenced by observations given by the FDA to several organizations, were presented.

The concept of process variation is an important topic in any manufacturing environment. Thus, understanding the difference between common causes of variation and special causes of variation, along with the basic principles of statistics, must be one of the first topics to include in any quality improvement endeavor. Our process knowledge can be enhanced by the use of graphical tools; many of them, along with specific examples of their application in an FDA-regulated organization, were presented throughout the book.

It is important to recognize that, prior to starting to collect information, we must make certain that our measurement system is reliable; that is, the measurement process is not adding more variation than the manufacturing process does. Once we have optimized our measurement system, then an assessment of the overall process variation versus the customer specifications must be performed in order to learn how capable our processes are. Then, hypothesis tests can help us understand the statistical differences between various groups, as well as the relationship between the variables that might have an impact on the process. However, no process improvement effort is comprehensive without the use of experimental design; that is, in order to find which are the factors that impact our key output variables (and which are the appropriate settings of those factors), some sort of systematic experimentation must be executed. It is important to recognize that experimentation is a continual process: with each additional experiment we will gain more process knowledge. Finally, but not less important, we need to continually monitor our processes, not only the process outputs but also the key process inputs. An excellent tool for continuously monitoring processes is the control chart. These charts must be used on a perpetual basis, not a "once-a-month" or "once-a-year" basis. Control charts will be one of the cornerstones of any process control system.

12.2 ORDER OF TOOLS

Throughout the book, many process improvement tools have been presented. I have made an attempt to present the tools in the specific order in which they must be implemented. Most of the improvement tools have been explained throughout the book; others have not been explained but can be found in many quality tools textbooks. What follows are, based on my experience, some of the recommended tools in any quality improvement effort, along with the order in which they must be applied.

Every improvement project must start with a *project charter.* This is a living document in which the details of the project are established. Topics such as project title, purpose, scope, goals, milestones, required resources, and so on, are agreed on by the person requesting the project and the person executing it. The charter must be updated as the project progresses. Once the project is defined and the team is organized, the *data measurement* process must begin. Some of the tools that can be used at this time are gage R&R, process capability analysis, histograms, box plots, dot plots, Pareto diagram, scatter plots, run charts, and others. What we are trying to accomplish in this part of the project is to gain some process understanding through the use of many graphical tools. Remember from Chapter 5 that each graphical tool has its specific objective.

As we progress through our process knowledge continuum, certain hypotheses are gradually developed. Graphical tools alone are not enough to prove those hypotheses; some sort of analytical evaluation must be performed. Hypothesis tests such as normality, equality of means, equality of medians, equality of variances, correlation between variables, statistical significance of factors, and so on, are appropriate at this stage. Just remember that most of these hypotheses are developed with already collected data. Because of that fact, the next logical step is to experiment in a systematic way. Recall from Chapter 10 that experimentation is sequential; do not expect to solve all your issues with a single experiment. Be prepared to develop and perform many experimental designs to achieve a better understanding of the process.

And remember, throughout the project (not only at the end of it), monitor your process with graphical tools such as the control chart. Realize that control charts are not only used to decide when to stop your process and take some action; control charts are the heart of any continuous quality improvement endeavor.

12.3 CONTINUOUS PROCESS MONITORING VERSUS ONCE-A-YEAR ANALYSIS AND REPORTING

Most pharmaceutical companies that I have consulted for during the past few years collect some sort of quality-related data on a lot-by-lot basis. However, just a few of them use the data to make decisions on a timely basis. My opinion is that those companies are reading the regulations in a literal manner. For instance, section 211.180 of 21 CFR §211 establishes the need for an *annual product review* (APR) for pharmaceutical companies. Specifically, it states that:

> Written records required by this part shall be maintained so that data therein can be used for evaluating, *at least annually*, the quality standards of each drug product to determine the need for changes in drug product specifications or manufacturing or control procedures.[1]

As noted in 21 CFR §211.180(e), the pharmaceutical regulations establish that records shall be analyzed *at least annually*. However, as mentioned in Section 1.3, while regulations are legally enforceable, guidances represent the agency's current thinking on certain topics. They serve to fill the gaps in the interpretation of the regulations. Particularly, the *Guidance for Industry: Quality System Approach to Pharmaceutical Current Good Manufacturing Practices* establishes:

> Quality systems call for continually monitoring trends and improving systems. This can be achieved by monitoring data and information, identifying and resolving problems, and anticipating and preventing problems.
>
> Quality systems procedures involve collecting data from monitoring, measurement, complaint handling, or other activities, and tracking this data over time, as appropriate. Analysis of data

> can provide indications that controls are losing effectiveness. The information generated will be essential to achieving problem resolution or problem prevention (see IV.D.3.).
>
> *Although the CGMP regulations (§211.180(e)) require product review on at least an annual basis, a quality systems approach calls for trending on a more frequent basis as determined by risk* [*emphasis added*]. Trending enables the detection of potential problems as early as possible to plan corrective and preventive actions. Another important concept of modern quality systems is the use of trending to examine processes as a whole; this is consistent with the annual review approach. Trending analyses can help focus internal audits (see IV.D.2.).[2]

Also, recall from Section 1.2 that regulations are the minimum requirements; that is, although the regulations establish at least a yearly basis for data analysis and process improvement, we must be more stringent than what the regulation establishes. In essence, we want to switch from a reactive mode to a proactive mode in order to prevent problems before they occur.

12.4 PROACTIVE OR REACTIVE?

As mentioned earlier, if we want to continuously improve our processes, we need to change one of the biggest paradigms we face every day: "If it's not broken, don't fix it." Oftentimes, people do not react until it is too late. As established in Section 3.1, we need to stop thinking that as long as our process is within the specification limits, nothing has to be done. Let me illustrate this concept with an example:

> A company is gathering data for their APR to determine whether changes have to be made to their process controls. One of the key process variables they measure is tablet hardness. The specification for that variable ranges from 4.0 to 11.0 kp. A control chart is developed in order to understand how that key process variable performed during the previous year. Figure 12.1 shows the individuals control chart for tablet hardness.

The argument presented in Figure 12.1 is the following: should we take action on those points outside of the control limits? Some people might say yes, while some people might say no. Those people that say some action must be taken will probably do so based on what we discussed in Chapter 11. Whenever a data point is outside of the control limits (or the chart

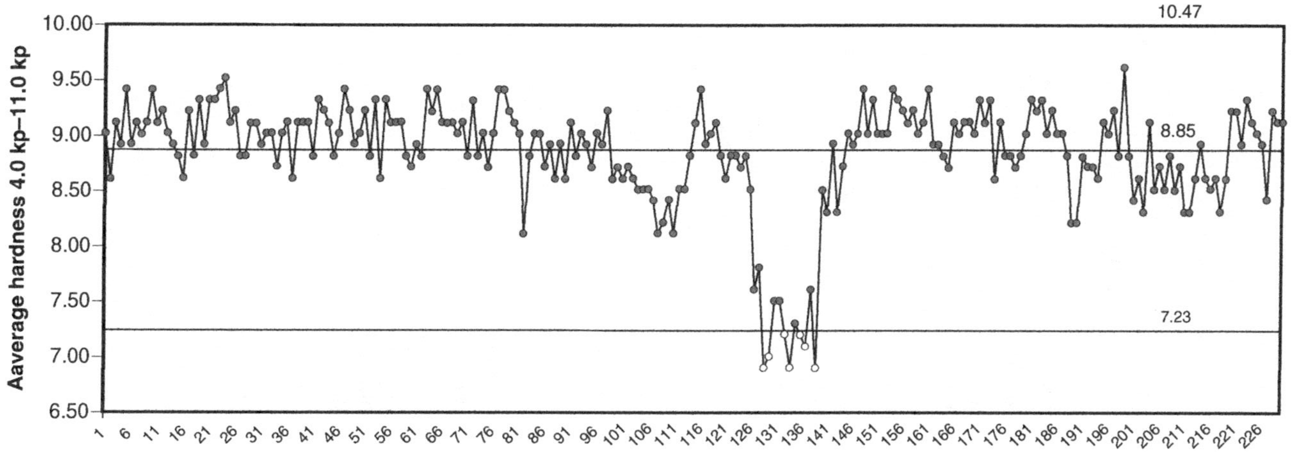

Figure 12.1 Individuals control chart for tablet hardness example.

is exhibiting a nonrandom pattern), that situation is being generated by an assignable, or special, cause. As mentioned, a special cause is something that must be investigated because it is not the inherent variation of the process (common cause) that is acting on the process, but a situation caused by some external factor. However, there will be other people who might say that since the process is still within the specification limits, nothing has to be done yet.

The situation presented above clearly demostrates the difference between two ways of thinking: *proactive* and *reactive*. As mentioned, if we want to continuously improve our processes, we need to switch from a reactive mode to a proactive mode. Before going further, let us see what the regulations establish on this topic. For instance, the regulation related to medical devices states that:

> (a) Each manufacturer shall establish and maintain procedures for implementing *corrective and preventive* action. The procedures shall include requirements for:
>
> (1) Analyzing processes, work operations, concessions, quality audit reports, quality records, service records, complaints, returned product, and other sources of quality data to identify *existing and potential* causes of nonconforming product, or other quality problems. Appropriate statistical methodology shall be employed where necessary to detect recurring quality problems;[3]

It can be noted that the medical device regulation explicitly mentions the need for corrective and *preventive* actions. Furthermore, it mentions the need to identify existing and *potential* causes of nonconforming product. So, if we take a look at Figure 12.1, it is true that it is not showing an existing nonconformance because it is still within the specification limits. However, it shows a potential nonconformance that must be addressed before it is too late. A reactive company would not do anything at this point, while a proactive company would investigate to find the cause of that potential nonconformance before a failure occurs. Which type of company is yours—reactive or proactive?

We have analyzed the regulation concerning medical devices. However, what happens in the finished pharmaceutical products arena? The regulation for finished pharmaceutical products establishes that:

> All drug product production and control records, including those for packaging and labeling, shall be reviewed and approved by the quality control unit to determine compliance with all established, approved written procedures before a batch is released or

> distributed. *Any unexplained discrepancy* (including a percentage of theoretical yield exceeding the maximum or minimum percentages established in master production and control records) or the failure of a batch or any of its components to meet any of its specifications *shall be thoroughly investigated*, whether or not the batch has already been distributed. The investigation shall extend to other batches of the same drug product and other drug products that may have been associated with the specific failure or discrepancy. A written record of the investigation shall be made and shall include the conclusions and follow-up.[4]

It is clear that the regulation on finished pharmeceutical products establishes that "any unexplained discrepancy must be thoroughly investigated." So, unless your company knows the causes for those out-of-control data points and has taken action to eliminate the possibility of those causes acting again in the process, an investigation must be enforced. Again, a reactive company would not do anything at this point, while a proactive company would investigate to find the cause of that unexplained discrepancy before a failure occurs. Which type of company is yours—reactive or proactive? In any case (medical devices or finished pharmaceutical products), what do we call the actions taken to avoid this pattern repeating again in the future—corrective actions or preventive actions? In essence, those would have to be called *preventive actions* because no failure has occurred yet. More information about the difference between corrective and preventive actions can be found in the book *CAPA for the FDA-Regulated Industry*, by José Rodríguez-Pérez.[5]

12.5 NEXT STEPS

Now that I have stressed the importance of being proactive instead of reactive, it is time to begin our journey through the quality improvement of our processes. I hope this book has fulfilled your expectations. As mentioned in the Preface, my goal was not to teach an intensive course in statistics, but to provide a how-to guide for the application of the diverse array of statistical tools available to analyze and improve the processes in an organization regulated by FDA. I hope that through your reading of this book you have obtained a better understanding of some of the available statistical tools for controlling the processes in your organization. Finally, I encourage you to study, with a greater level of detail, each of the statistical tools presented throughout the book.

Appendix A
Variable and Attribute Data Applications

Many times, we are not certain about which type of tool to apply for a specific situation. For instance, we want to develop a control chart but do not know which one is the most appropriate for the type of data at hand. The same doubts could be generated when deciding which distribution to use for the analysis of data. The following table shows a non-exhaustive list of some of the tools applicable to variable or attribute data.

Table A.1 Variable and attribute data applications.

	Variable data	**Attribute data**
Characteristics	Measurable Continuous	Counted Discrete Categories Binary
Examples	Temperature Length Time Speed	Number of defects Percent defective units Pass/fail Go/no-go
Control charts	$\bar{X}$ and R $\bar{X}$ and s ImR Medians chart	*p*-chart *np*-chart *c*-chart *u*-chart
Distributions	Normal Exponential Weibull Lognormal	Poisson Binomial Hypergeometric
Sampling plans	ANSI/ASQ Z1.9	ANSI/ASQ Z1.4
Measurement instruments	Caliper Micrometer Scales	Plug gage Ring gage Pin gage

Appendix B
Applications for Various Graphical and Statistical Tools

Many times, we are faced with the following dilemma: we want to analyze some data but do not know which tool to use in order to analyze the data. In this appendix, I will split the discussion into two topics: graphical tools and statistical tools. The following tables present a non-exhaustive list of the tools available for the kind of analysis we want to perform.

Table B.1 Applications for graphical tools.

If you want to . . .	Then you can use . . .	Find tool in Section
Visualize the shape, central tendency, and dispersion of data	Histogram	5.2
Compare central tendency and dispersion of various groups	Box plot	5.3
Prioritize the order in which you deal with certain issues	Pareto diagram	5.5
Show the relationship between an independent variable and a dependant variable	Scatter plot	5.6
Identify the common and/or special causes in a process	Control chart	11.1
Determine whether data are random or exhibit some trend	Run chart	5.7

Table B.2 Applications for statistical tools.

If you want to . . .	Then you can use . . .	Find tool in Section
Learn about the measures of central tendency, dispersion, and shape of the data	Descriptive statistics	4.4
Evaluate the normality of the data	Anderson-Darling normality test	5.8
Identify the variation caused by the process, by the measurement instrument, and by the analysts	Measurement systems analysis (gage R & R)	6.1
Compare the process variation against the specification limits	Process capability analysis	7.1
Compare an average to a single value	One-sample *t*-test	8.2.1
Compare the averages of two groups	Two-sample *t*-test	8.2.2
Compare the averages of two or more groups	ANOVA	8.2.3
Compare a median to a single value	One-sample sign test	8.3.1
Compare the medians of two groups	Two-sample Mann-Whitney test	8.3.2
Compare the medians of two or more groups	Kruskal-Wallis test	8.3.3
Compare the variances of two groups	*F*-test	8.4.1
Compare the variances of three or more groups	Bartlett test	8.4.2
Compare the variances of nonnormal data	Levene test	8.4.3
Evaluate the relationship between an independent variable and a dependent variable	Simple linear regression	9.5
Evaluate the relationship between two or more independent variables and a dependent variable	Multiple linear regression	9.6
Identify the input variables that have an impact on the output variables	Design of experiments	10.1

Table B.2 *Continued.*

If you want to . . .	Then you can use . . .	Find tool in Section
Identify sources of process variation for variable data (subgroup size = 1)	ImR chart	11.5.1
Identify sources of process variation for variable data (subgroup size from 2 to 6)	$\bar{X}$ and R chart	11.5.2
Identify sources of process variation for variable data (subgroup size > 6)	$\bar{X}$ and s chart	11.5.3
Identify sources of variation for attribute data (percent defective)	*p*-chart	11.6.1
Identify sources of variation for attribute data (number of defectives)	*np*-chart	11.6.2
Identify sources of variation for attribute data (number of defects)	*c*-chart	11.6.3
Identify sources of variation for attribute data (number of defects per unit)	*u*-chart	11.6.4

Appendix C
Basic Statistics for an Annual Product Review (APR) Report

Section 211.180 of 21 CFR §211 establishes the need for an annual product review (APR) for pharmaceutical companies. Specifically, 21 CFR §211.180(e) states that:

> Written records required by this part shall be maintained so that data therein can be used for evaluating, *at least annually*, the quality standards of each drug product to determine the need for changes in drug product specifications or manufacturing or control procedures.[1]

In the medical device arena, section 820.20 of 21 CFR establishes the need for management review. Specifically, 21 CFR §820.20(c) states:

> Management Review. Management with executive responsibility shall review the suitability and effectiveness of the quality system at defined intervals and with sufficient frequency according to established procedures to ensure that the quality system satisfies the requirements of this part and the manufacturer's established quality policy and objectives. The dates and results of quality system reviews shall be documented.[2]

So, based on the regulations presented above, there is a need to evaluate the quality standards and quality system in order to evaluate their effectiveness and determine the need for changes and improvement. The regulations do not establish which type of analysis will have to be performed to achieve these goals. In the next example, I present a systematic analysis for one parameter (assay) and the graphical and statistical tools recommended to analyze that parameter in order to comply with the requirements of the regulations.

The analysis starts with a descriptive statistics study. Measures of central tendency (mean and median), measures of dispersion (range, standard deviation), confidence intervals, and normality tests are some of the statistics analyzed. Then, a process capability analysis is performed to determine how capable the process is to produce within the specification limits. Also, an individuals and moving range chart is used to evaluate whether the process is in statistical control for the year for which the annual product review is being developed.

Then, another individuals and moving range chart is used to compare the current year with the previous year to see if there has been any significant change in the central tendency or dispersion of the process. To complement the control chart, and to determine in a statistical way if the change in central tendency and dispersion seen in the control chart is significant or not, an ANOVA and Bartlett test are performed. Let us see the details of the report:

ANNUAL PRODUCT REVIEW FOR AMIPLINATO TABLETS, 250 MG—JANUARY 2012–DECEMBER 2012

PARAMETER: ASSAY (SPEC = 90.0%–110.0%)

Results from assay analysis (Figure C.1) indicate that this parameter follows a normal distribution with an average of 98.95% and a range between 95.90% to 101.60%, well within specifications. This parameter fits a normal distribution with a *p*-value for the Anderson-Darling normality test

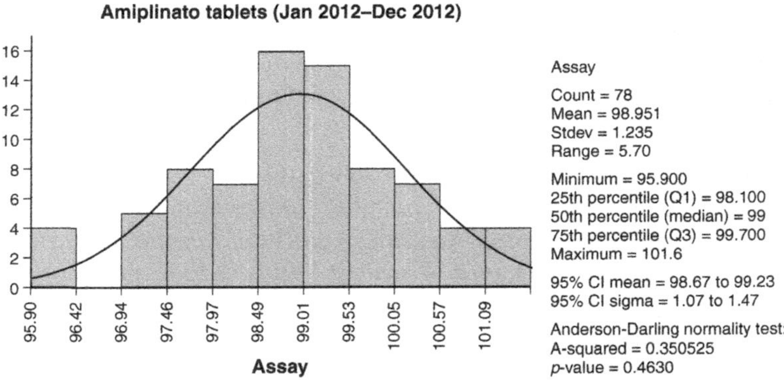

Figure C.1 Graphical summary for Amiplinato tablets APR.

of 0.4630. A p-value lower than 0.05 would indicate that the data do not fit a normal distribution. The 95% confidence interval for the mean ranges from 98.67% to 99.23%, showing the robustness of this parameter when compared to the specification limits.

The same indication of robustness is obtained from the analysis of the process capability indices (Figure C.2), with a C_p value of 3.15 and a C_{pk} of 2.82. A C_p value higher than 1.33 is considered acceptable as a measure of process capability.

The individuals and moving range chart (Figure C.3) depicts sequentially all batches manufactured during the period of January 2012 to December 2012. The behavior of this parameter seems to be very stable, with minimal batch-to-batch variations. Despite such variation, all the individual datapoints are well within the specification limits. Consequently, no further action is recommended.

Previous Year Comparison

In order to understand the behavior of this parameter when compared with the previous year, an individuals chart was developed. The individuals chart (Figure C.4) shows the average, upper control limit, and lower control limit for the periods of January 2011 to December 2011 and January 2012 to December 2012. A slight increase in the process centering can be observed for the period of January 2012 to December 2012. The statistical significance of these results will be analyzed through an analysis of variance (ANOVA) test and Bartlett test.

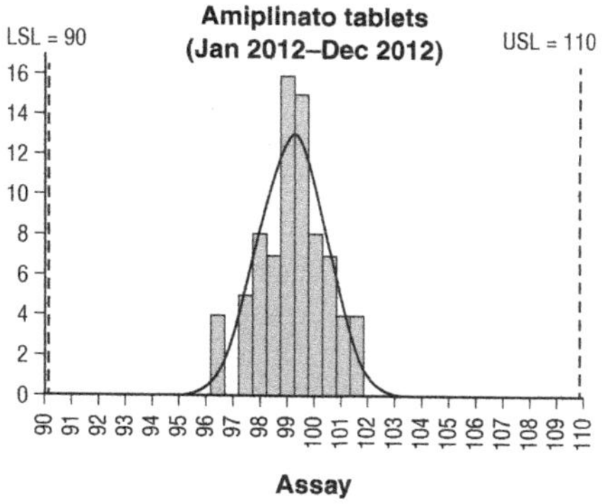

Process capability report: Assay	
Count	78
Mean	98.951
StDev (overall, long term)	1.235
StDev (within, short term)	1.058073
USL	110
Target	
LSL	90

Potential capability indices using within StDev	
Cp	3.15
Cpu	3.48
Cpl	2.82
Cpk	2.82
Cpm	

Figure C.2 Process capability analysis for Amiplinato tablets APR.

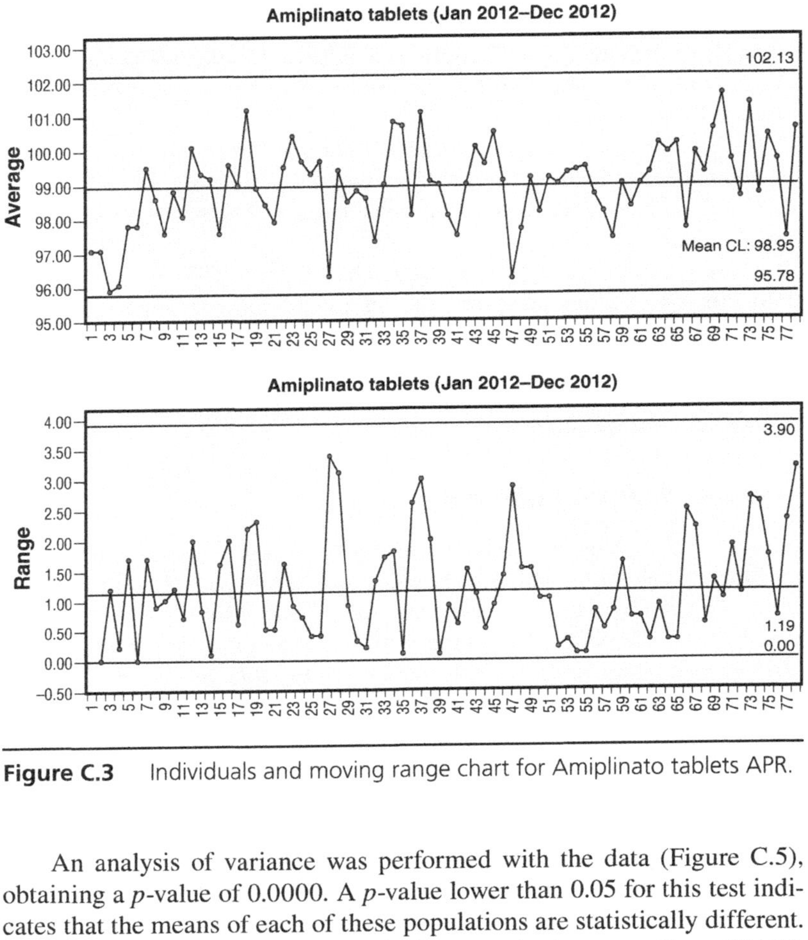

Figure C.3 Individuals and moving range chart for Amiplinato tablets APR.

An analysis of variance was performed with the data (Figure C.5), obtaining a *p*-value of 0.0000. A *p*-value lower than 0.05 for this test indicates that the means of each of these populations are statistically different. The obtained value indicates that the increase in the mean (from 97.57% for the period of January 2011 to December 2011 to 98.95% for the period of January 2012 to December 2012) is due to some special cause(s). This increase in the assay mean is not the result of natural process variability. An investigation is recommended to find the causes of the increase in the assay mean from one year to the next.

The standard deviations for each year are as follows: 1.113 for the period of January 2011 to December 2011, and 1.235 for the period of January 2012 to December 2012. Performing a Bartlett analysis of the data (Figure C.6), a *p*-value of 0.4338 was obtained. A *p*-value higher than 0.05 for this test indicates that no significant differences are observed between the variances of each of these populations. It can be concluded that the

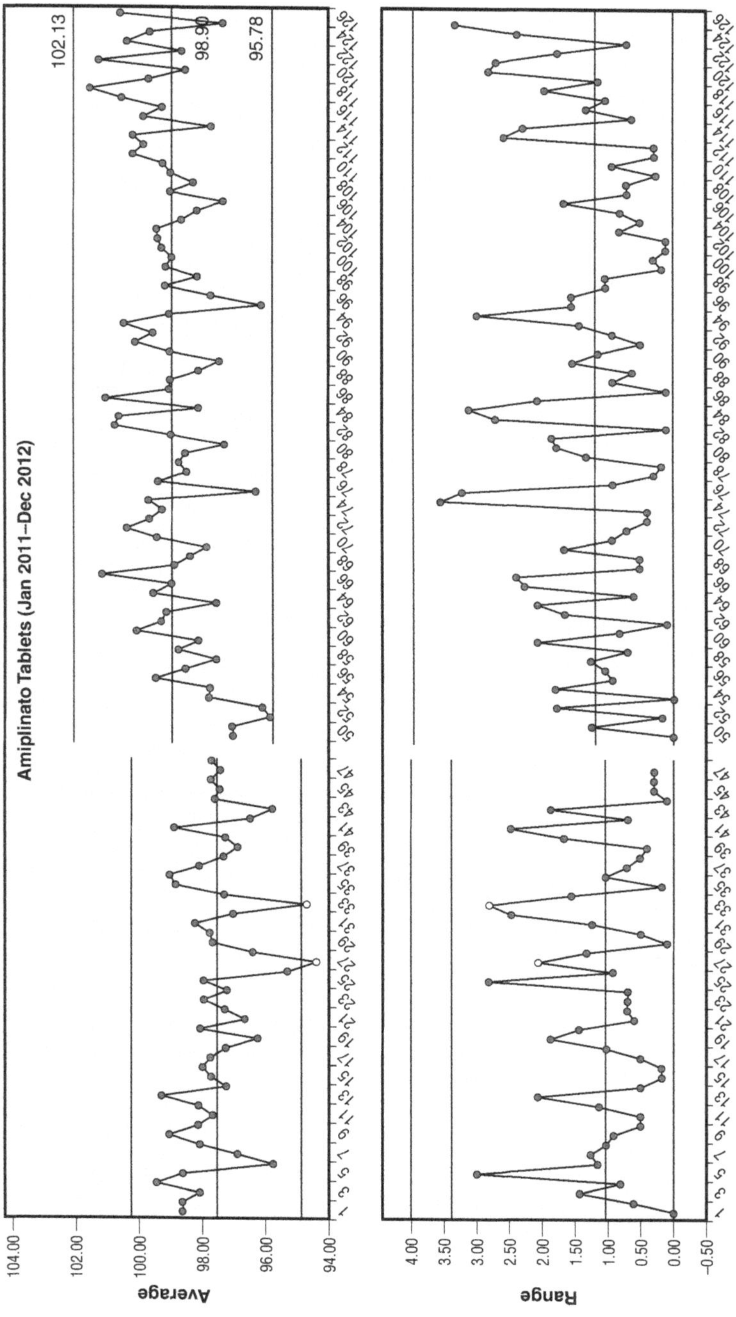

Figure C.4 Two-year comparison for Amiplinato tablets APR.

One-way ANOVA and means matrix: assay

Test information

H_0: Mean 1 = Mean 2 = . . . = Mean *k*

H_a: At least one pair Mean *i* ≠ Mean *j*

Year	**Jan 2011–Dec 2011**	**Jan 2012–Dec 2012**
Count	48	78
Mean	97.565	98.951
StDev	1.113	1.235
UC (2-sided, 95%, pooled)	97.905	99.218
LC (2-sided, 95%, pooled)	97.225	98.685

ANOVA table

Source	**SS**	**DF**	**MS**	***F***	***p*-value**
Between	57.139	1	57.139	40.352	**0.0000**
Within	175.58	124	1.416		
Total	232.72	125			

Figure C.5 ANOVA test for Amiplinato tablets APR.

Bartlett's test for equal variance: assay
(Use with normal data)

Test information

H_0: Variance 1 = Variance 2 = . . . = Variance *k*

H_a: At least one pair Variance *i* ≠ Variance *j*

Year	**Jan 2011–Dec 2011**	**Jan 2012–Dec 2012**
Count	48	78
Mean	97.565	98.951
Median	97.750	99
StDev	1.113	1.235
AD normality test *p*-value	**0.0281**	0.4630
Bartlett's test statistic	0.612510	
p-value	0.4338	

Figure C.6 Bartlett test for Amiplinato tablets APR.

observed year-to-year variation is not significantly different. Consequently, no further action is recommended.

The overall conclusion when analyzing results from batches of Amiplinato tablets produced during the period of January 2012 to December 2012 is that all analyzed parameters behave well within specifications.

The process capability indices were far above 1.00 in all cases, indicating that the process is capable of producing within the specification limits. Slight differences were detected in some of the parameters when compared year to year. However, when the means were compared (using ANOVA) and variances were compared (using the Bartlett test), these differences were not statistically significant. In summary, the Amiplinato tablets produced during the period of January 2012 to December 2012 complied with all parameters, and no further changes in specifications or process controls are required.

Appendix D
Most Commonly Used Hypothesis Tests

At times, we do not know which hypothesis test to apply in a certain situation. Table D.1 shows the most commonly used hypothesis tests and how to interpret each one. The analysis will be based on an obtained *p*-value from any statistical software package.

Table D.1 Most commonly used hypothesis tests.

Test	Null and alternate hypothesis
General	H_0: Groups are not significantly different H_a: Groups are significantly different
Normality	H_0: Data follow the normal distribution H_a: Data do not follow the normal distribution
Randomness of data	H_0: Data are random H_a: Data are not random
Two-sample *t*-test	H_0: Means are not significantly different H_a: Means are significantly different
ANOVA test	H_0: Means are not significantly different H_a: At least one mean is significantly different
F-test	H_0: Variances are not significantly different H_a: Variances are significantly different
Two-proportions test	H_0: Proportions are not significantly different H_a: Proportions are significantly different
Regression	H_0: Data are not correlated H_a: Data are correlated
Design of experiments	H_0: Effect is not significant H_a: Effect is significant

How to interpret this table:

- If the p-value $< \alpha$, reject H_0 and accept H_a.
- If the p-value $\geq \alpha$, there is no evidence to reject H_0.

Endnotes

Chapter 1

1. "What Does FDA Do?" FDA (Food and Drug Administration), last modified December 17, 2010, accessed January 17, 2013, http://www.fda.gov/AboutFDA/Transparency/Basics/ucm194877.htm.
2. 21 CFR §1.1 to §1499.
3. "What Is the Difference between the Federal Food, Drug, and Cosmetic Act (FD&C Act), FDA Regulations, and FDA Guidance?" FDA, last modified August 19, 2010, accessed January 7, 2013, http://www.fda.gov/AboutFDA/Transparency/Basics/ucm194909.htm.
4. 21 CFR §211.1 to §211.208.
5. 21 CFR §820.1 to §820.250.
6. 21 CFR §211.1(a).
7. 21 CFR §820.1 (a)(1).
8. 21 CFR §211.22.
9. 21 CFR §211.100.
10. 21 CFR §820.250.
11. 21 CFR §820.100.
12. Food and Drug Administration, *FDA Guidance for Industry: Quality System Approach to Pharmaceutical Current Good Manufacturing Practices*, section IV.C.3 "Perform and Monitor Operations" (Washington, D.C.: FDA, 2006), 19.
13. Food and Drug Administration, *FDA Guidance for Industry: Investigating Out-of-Specification (OOS) Test Results for Pharmaceutical Production*, section IV.C.1(a) "Reporting Testing Results: Averaging; Appropriate Uses" (Washington, D.C.: FDA, 2006), 9.
14. 21 USC 351(a)(2)(B).
15. Food and Drug Administration, *Guidance for Industry: Process Validation; General Principles and Practices*, section IV.B.2 "Establishing a Strategy for Process Control" (Washington, D.C.: FDA, 2011), 9.

16. José Rodríguez-Pérez, *CAPA for the FDA-Regulated Industry* (Milwaukee: ASQ Quality Press, 2011), 25.
17. ICH Q10 Pharmaceutical Quality System, section 3.2, "Pharmaceutical Quality System Elements," 7.
18. Rodríguez-Pérez, *CAPA*, 26.
19. International Organization for Standardization, ISO 13485:2003 *Medical devices—Quality management systems—Requirements for regulatory purposes*; section 8.4, "Analysis of data" (Geneva: ISO, 2003).

Chapter 2

1. "Inspections, Compliance, Enforcement, and Criminal Investigations: Merge Healthcare 8/24/12," FDA, last modified August 29, 2012, accessed January 8, 2013, http://www.fda.gov/ICECI/EnforcementActions/WarningLetters/2012/ucm317269.htm.
2. "Inspections, Compliance, Enforcement, and Criminal Investigations: Selder S.A. de C.V. 4/27/12," FDA, last modified May 16, 2012, accessed January 8, 2013, http://www.fda.gov/ICECI/EnforcementActions/WarningLetters/2012/ucm303962.htm.
3. "Inspections, Compliance, Enforcement, and Criminal Investigations: Diagnostics Biochem Canada, Inc. 12/20/11," FDA, last modified April 16, 2012, accessed January 8, 2013, http://www.fda.gov/ICECI/EnforcementActions/WarningLetters/2011/ucm300296.htm.
4. "Inspections, Compliance, Enforcement, and Criminal Investigations: Sunrise Pharmaceutical, Inc. 1/14/10," FDA, last modified January 13, 2012, accessed January 8, 2013, http://www.fda.gov/ICECI/EnforcementActions/WarningLetters/2010/ucm197966.htm.
5. "Inspections, Compliance, Enforcement, and Criminal Investigations: Arrow International, Inc. 10-Oct-07," FDA, last modified July 7, 2009, accessed January 8, 2013, http://www.fda.gov/ICECI/EnforcementActions/WarningLetters/2007/ucm076532.htm.
6. "Inspections, Compliance, Enforcement, and Criminal Investigations: Gentere, Inc. 13-July-04," FDA, last modified July 8, 2009, accessed January 8, 2013, http://www.fda.gov/ICECI/EnforcementActions/WarningLetters/2004/ucm146503.htm.
7. "Inspections, Compliance, Enforcement, and Criminal Investigations: Alphatec Spine, Inc. 6/21/10," FDA, last modified October 26, 2011, accessed January 8, 2013, http://www.fda.gov/ICECI/EnforcementActions/WarningLetters/2010/ucm223696.htm.
8. 21 CFR §820.100.
9. Food and Drug Administration, *FDA Guidance for Industry: Investigating Out-of-Specification (OOS) Test Results for Pharmaceutical Production*, Section IV "Investigating OOS Test Results—Phase II: Full-Scale OOS Investigation" (Washington, D.C.: FDA, 2006), 6.

Chapter 3

1. Genichi Taguchi, Subir Chowdhury, and Yuin Wu, *Taguchi's Quality Engineering Handbook* (Hoboken, NJ: John Wiley & Sons, 2005).

Chapter 5

1. José Rodríguez-Pérez and Manuel E. Peña-Rodríguez, "Fail-Safe FMEA," *Quality Progress* (January 2012): 30–36.

Chapter 12

1. 21 CFR §211.180(e).
2. Food and Drug Administration, *FDA Guidance for Industry: Quality System Approach to Pharmaceutical Current Good Manufacturing Practices*, section IV.D.1 "Evaluation Activities; Analyze Data for Trends" (Washington, D.C.: FDA, 2006), 21.
3. 21 CFR §820.250.
4. 21 CFR §211.192.
5. José Rodríguez-Pérez, *CAPA for the FDA-Regulated Industry* (Milwaukee: ASQ Quality Press, 2011), 6–7.

Appendix C

1. 21 CFR §211.180(e).
2. 21 CFR §820.20(c).

Index

A

B

C

J

K

L

M

N

O

P

U

V

W

X